MARE Publication Series

Series editors
Maarten Bavinck
University of Amsterdam, Amsterdam, The Netherlands

Svein Jentoft
Tromsø, Norway

The MARE Publication Series is an initiative of the Centre for Maritime Research (MARE). MARE is an interdisciplinary social-science network devoted to studying the use and management of marine resources. It is based jointly at the University of Amsterdam and Wageningen University (www.marecentre.nl).

The MARE Publication Series addresses topics of contemporary relevance in the wide field of 'people and the sea'. It has a global scope and includes contributions from a wide range of social science disciplines as well as from applied sciences.

Topics range from fisheries, to integrated management, coastal tourism, and environmental conservation. The series was previously hosted by Amsterdam University Press and joined Springer in 2011. The MARE Publication Series is complemented by the Journal of Maritime Studies (MAST) and the biennial People and the Sea Conferences in Amsterdam.

Editors
Svein Jentoft
University of Tromsø, Norway
Svein.jentoft@uit.no

J. Maarten Bavinck
University of Amsterdam, The Netherlands
J.M.Bavinck@uva.nl

More information about this series at http://www.springer.com/series/10413

Jeppe Høst

Market-Based Fisheries Management

Private fish and captains of finance

 Springer

Jeppe Høst
Department of Ethnology Saxo Institute
University of Copenhagen
Copenhagen
Denmark

ISSN 2212-6260 ISSN 2212-6279 (electronic)
MARE Publication Series
ISBN 978-3-319-16431-1 ISBN 978-3-319-16432-8 (eBook)
DOI 10.1007/978-3-319-16432-8

Library of Congress Control Number: 2015934939

Springer Cham Heidelberg New York Dordrecht London
© Springer International Publishing Switzerland 2015
This work is subject to copyright. All rights are reserved by the Publisher, whether the whole or part of the material is concerned, specifically the rights of translation, reprinting, reuse of illustrations, recitation, broadcasting, reproduction on microfilms or in any other physical way, and transmission or information storage and retrieval, electronic adaptation, computer software, or by similar or dissimilar methodology now known or hereafter developed.
The use of general descriptive names, registered names, trademarks, service marks, etc. in this publication does not imply, even in the absence of a specific statement, that such names are exempt from the relevant protective laws and regulations and therefore free for general use.
The publisher, the authors and the editors are safe to assume that the advice and information in this book are believed to be true and accurate at the date of publication. Neither the publisher nor the authors or the editors give a warranty, express or implied, with respect to the material contained herein or for any errors or omissions that may have been made.

Printed on acid-free paper

Springer is part of Springer Science+Business Media (www.springer.com)

Preface

As I have navigated the difficult waters of market-based fisheries management, many people have helped and influenced my path. Privatization of fishery resources has continuously been on the political agenda in many countries and thus been the object of discussion in fishing communities, at academic conferences and parliamentary hearings as well as in public media. It has been my privilege to bridge and bind together many of these diverse and genuine debates. Thus, I am deeply grateful to the many people who have contributed to my work and journey. Although I am indebted to many, the ideas and conclusions presented here are my responsibility alone.

There are many people that I would like to thank. Surely, friends, family, and colleagues have been exposed to a lot of talk on "fish quotas," and your many questions have constantly led me to find new ways to explore and explain my arguments. Throughout the process, my supervisor Professor Thomas Højrup has shared his immeasurable knowledge and passion for the field, of which I am deeply indebted and inspired. In addition, I am very grateful for the financial support from the Ministry of Food, Agriculture and Fisheries, the University of Copenhagen and from the Research Council, as well as the many funds for traveling I have received from, among others, DANIDA, The European Parliament, OCEAN 2012, and Solarfonden. This book is based on my PhD thesis from 2013. Since then, the manuscript has been revised and improved thanks to a number of peer reviewers. Finally, my girlfriend Katrine has been an outstanding support and has ceaselessly shared her love and optimism.

There are other vessels out there navigating the waters of market-based fisheries and I have learned continuously from the cooperation with NGOs, community entrepreneurs, fishers, and other scholars from all over the world. I hope this work will propel and enlighten your quest as well. There are strong currents in the international environment favoring a privatization of the oceans. Step by step a far-reaching privatization of the ocean is being implemented with undesirable social consequences. *People say don't rock the boat, let things go their own way.* I say *let's rock the boat.*

Contents

1	Introduction	1
	Market-Based Fisheries Management	1
	Fisheries Management as a Social Science	2
	A Brief Introduction to Market-Based Fisheries Management	4
	A Disciplinary Gap	5
	The Other Side of Concentration	6
	A Neoliberal Turn?	8
	Neoliberalism of the Oceans	8
	References	11
2	Growth and Management	13
	Long Past, Short History	13
	Sharing	16
	Technology at Sea	18
	Harbors and Land-Based Infrastructure	20
	Markets, Demand, and Distribution	21
	Globalization and Competition	23
	Cannons and Commons	23
	Expanding EEZs	24
	State, Growth, and Marshall Funding	26
	Organizational Fragmentation	28
	Small Craft and Trash Fish	29
	The Introduction of Output Management	30
	The Quota Distribution Problem	33
	Two Decades of Rations	34
	Diverging Solutions, Groups, and Interests	35
	The Arrival of Market-Based Fisheries Management	36
	Captains of Finance and the New Regulation	37
	A Break with Equal Access	39
	Markets—the Problem and the Solution?	40
	References	41

3	**Society and Market**	45
	Market Economy	45
	Division of Labor and the Economic Man	46
	Reciprocity, Redistribution, and Householding	47
	The Market Economy and Social Safeguards	49
	The Market Economy and Fishing Rights	50
	The Market Economy and Social Science	51
	Embedded, Disembedded, or Both?	52
	Safeguarding and Market-Based Fisheries	54
	The Economics of Market-Based Fisheries Management	55
	Multiple Objectives	56
	The State and Free Markets	56
	The Commodity and Principle of Catch History	58
	The Rights Holder	59
	Transactions	60
	Concentration and Maximum Ownership	60
	Concentration and Average VQS Size	64
	Fleet Segmentation and the Coastal Fishery Safeguard	66
	Transfers Between Fleets Segments	68
	Quota Blocks and Gear Differentiation	70
	Geography: Area Specific Shares	72
	Fewer Boats and Empty Harbors	73
	Markets in Motion	75
	Market-Based Transformation	76
	References	78
4	**The Commodity and its Exchange**	81
	The VQS Commodity	81
	The Commodity as the Framework	82
	The Commodity	83
	Use and Exchange Values	83
	The Two Aspects of the Rights Holder	85
	TAC and VQS	85
	Fluctuations and New Opportunities	86
	Exchange and the Marketplace	87
	Leasing	88
	Exchange Situations	89
	Ethnographies of Exchange	90
	About Selling	90
	The Sale	93
	To Buy or Not to Buy?	94
	Leasing or Losing?	97
	Value and Value Fluctuations	100
	Labor or Rent	101
	Monopoly Rent	103

	Monopoly Fisheries?	104
	Conclusions	106
	References	107
5	**Access and Fishing Activities**	109
	Heterogeneity	110
	The One-Man Operation	110
	Direct Sales and Tourists	113
	Young New Entrant	114
	Fishing Strategy	115
	The Vessel and the Fishing Activities	116
	Thrift and Flexibility	118
	The Tenant	118
	The Vessel and the Future	120
	The New Company	121
	The Vessel and Fishing Activities	122
	Production and Landings	123
	The Large-Scale Operation	125
	Fishing Activity and Strategy	126
	Labor Organization	127
	The Great Expansion	128
	Discussion: What Is New in the VQS System?	128
	General Patterns	131
	Expansive Restructuring	132
	Balanced Readjustment	133
	Investment Aversion	134
6	**Transformation and Modes of Production**	137
	Introduction	137
	International or National Agenda	139
	Two Modes of Production	139
	Blank Sheet	140
	Simple Commodity Mode of Production	141
	The Commodity Market	142
	Fixed and Variable Costs	143
	Price Manipulation	144
	Manipulation of Costs	145
	Quantity Manipulation	145
	The Catch Unit Further Specified	146
	Maximum Production	146
	Switching Fishery	147
	Normal Production	147
	Vessel Circulation	148
	Constant Adaptations and the Share System	148
	Ideological Relations	149

	Discussion: Principal Changes with the VQS System	151
	Set Quantity as an Entry Point	151
	Crew and Share Organization	152
	Withering Away	153
	Capitalist Fisheries	154
	Isafold	156
	Limits to Growth	157
	Organization on Board	157
	The VQS and the Capitalist Mode of Production	159
	Crew and New Relations	160
	Life-Modes of the Sea	160
	Conclusions	162
	References	163
7	**Postscript: Everyday Life and Mediated Fisheries**	165
	The Economics of Ethnology	165
	A New Everyday	166
	European Union and Beyond	168
	Last Generation of Self-Employed Fishers	169
	Mediated Fisheries	170
	Private Fish and Captains of Finance	171
	References	171

Chapter 1
Introduction

Abstract This chapter introduces the notion and theme of market-based fisheries management. The introduction of market mechanisms to distribute and manage fishing quota has internationally occurred since the 1980s but is increasingly on the political agenda. As privatization and transferability are promoted internationally, and by big players, it becomes even more crucial to understand its social and environmental implications. This chapter looks at the broader lines in the fisheries research and discusses the diverging views on market-based fisheries management. On one side market mechanisms are promoted for efficiency and fleet adaptation, and on the other, quota markets are accused of causing concentration and disturbing social relations.

Keywords Individual transferable quotas · Neoliberalism · Social anthropology · Fisheries management · Overcapacity · Ocean commons

Market-Based Fisheries Management

The main objective of this book is to examine the 2007 introduction of a market-based fisheries management system in Denmark. The introduction of individual, transferable and privately owned fishing quotas led to vast changes in daily practices and demanded whole new strategies with which to approach the practice of fishing. New objects like "lease-fish" have entered the vocabulary of fishers, together with new opportunities and new challenges. In a market-based fisheries management system, limited fishing opportunities are distributed through market mechanisms. Vessel owners can buy or lease "quota"—the right to catch a share of the nation's fish—from other vessel owners. To enter the fisheries, one has to acquire "fish" either by buying or leasing fishing quota. This is a radical break from the previous management systems, within which principles of free and equal access prevailed and where the resource was still state owned. The market-based fisheries management systems I study in this book shares many common characteristics with other market-based systems both in Denmark and internationally. With the term market-based fisheries management, I refer to systems known under the categories of individual transferable quotas, transferable fishing concessions, catch shares, individual fishery quotas, wealth-based management and others. Individual cases of the market-based model have many names, but all share the same characteristic of a market to distribute fishing opportunities which in turn replaces prior management principles including, for example, licenses, open access or limited entry.

Ownership of quota gives fishers the right to a certain share of the mandated national landings of a specific species in a specific catch area. In a market-based management system, the distribution of fishing opportunities—a core management issue—is handled by economic transactions in a market and not, as previously, by the state. In the following chapters, I try to uncover how such a system works on an everyday level. I aim to understand and explain market-based fisheries management as an everyday phenomenon. I will ask what a market is and what forms the relation between society, social groups and the market in question. How do the actions undertaken on this market influence the ownership structure of the resource, and how does "quota" function as a commodity? What is it worth? How does it influence fishing operations and change fishing as an occupation, from the practical day-to-day planning to the lifelong career perspectives of individual fishers? These questions are investigated through a study of the Vessel Quota Share (VQS) system, a multi-species market-based management system introduced in 2007, to manage the distribution of commercial demersal fisheries in Denmark. Implemented in Denmark simultaneously for a wide range of species and covering all catch areas around Denmark, the Danish adaptation of market-based fisheries management is perhaps one of the most extensive and far-reaching of its kind. In essence, the Danish VQS system attaches a fixed percentage of the national quota to each boat[1]. The basic principle, and the most novel feature of this system, is that boat owners can buy another boat in order to acquire its quota and fishing rights. In principle, the quota is therefore transferred through acquisition of boats. However, there is flexibility in the system that allows the quota and boat to be separated from one another and the quota traded without the vessel it was originally attached to. This process is further described in Chap 3, "Society and Market".

Fisheries Management as a Social Science

Why should an ethnologist be concerned with fisheries management? Is it not an arena reserved for marine biologists, economists and other specialists on resource management? The short answer is that fisheries management is not just about fish; it is also about the people and therefore also a social and cultural issue. We can

---

[1] Initially, this allocation was based on a 3-year catch history of demersal species. Species managed by the VQS system are cod, sole, plaice, Norwegian lobster, coalfish, haddock and deep sea prawns in all catch areas; hake and turbot in the North Sea; monkfish in Norwegian territory; the demersal fishery of sprat; and herring and salmon in the Baltic Sea. Some pelagic fisheries in Denmark (herring, sprat and sand eel) have already been managed since 2003 by ordinary ITQ systems.

only manage fish stocks through managing human behaviour. As a result, fisheries management is primarily concerned with the human use of marine resources. Fisheries management is the management of fish stocks through the management of the people who catch the fish. Market-based fisheries management in particular is based on interpersonal transactions. These diverse economic transactions are rooted in fishing operations that consist of people working together. Some of these companies are new, and some are made up of crews that have worked together for many years. These business transactions are attached to personal meanings and values. Some economic transactions occur without any profit motive, as for example when a local owner leases quota at a discount rate to a young, newly-qualified skipper. Some are concerned purely with the profit that can be accrued from fluctuations in quota prices. Therefore, social science broadly understood, has an important place in the analysis of fishing communities and fishing practices under market-based fisheries management.

Fisheries management has, until recently, never been a distinct research subject for anthropologists and ethnologists. However, in response to resource management discussions about the "tragedy of the commons" (Hardin 1968), geographers, historians, anthropologists and ethnologists have identified and documented social institutions governing common resources (McCay 2001; Ostrom 2002). Engaging in this discussion was a step towards recognizing management as a distinct research subject. In this way, scholars could draw on both earlier work and the more recent community studies of the 1960s and 1970s. Examples from Scandinavian ethnology include the works of Holger Rasmussen (1968) and Olof Hasslöf (1949). Both give evidence that historical fishing activities were neither unmanaged nor unorganized, and that intensive resource extraction is as much about continuity as it is about radical growth in the latter half of the twentieth century. In this context, fisheries management is embedded in social institutions and structures in the communities and between people—i.e. the resource as common property or access structured through social relations—and not understood as the scientific alliance of biology and economy that we are familiar with today. Even in the community studies where regulation is sometimes present as severe restrictions on activities, these are considered temporary, external and not an object of study (Moustgaard and Damgaard 1974). So-called scientific management of fishing activities grew in both reach and importance in the second half of the twentieth century, and increasingly became entrenched in state bureaucracy as well as international politics. In cultural studies, this alliance of science and government has been the object of institutional analysis (see for an example Charles 2001) and has been mapped out as part of heterogeneous techno-social networks (see for example Holm 2001). In the last decades, scholars have documented the implications of fisheries management on communities and local conditions and on rural development (Brox et al. 2006; McCay 2001). Regulations, policies and fisheries management in a broad sense, therefore, now constitute a complex research area with multiple disciplines involved.

A Brief Introduction to Market-Based Fisheries Management

Internationally, market-based fisheries management was first advocated in the 1970s as a response to the increasing problem of growing fishing capacity (McCay 1995). Its actual development in Danish fisheries will be examined in Chap. 2, "Growth and Management". What is worth noting here is that market-based fisheries management was, in the early days of its development, just one option among the many put forward to solve the increasing pressure on fish populations and the problematic distribution of limited resources. Privatization was controversial and did not come about without resistance. It is difficult to know why neoliberal views on the privatization of fish resources, once considered marginal, are now, a few decades later, considered mainstream. The answer to this question lies in diverse areas. Shifts in state bureaucracies, where economists and economic models have gained power and influence, might be part of the explanation, as are globalized markets and harsh competition from the outside. Since the 1980s, the Western world has witnessed a general deregulation of trade and financial assets, which might have influenced the way fishery resources are viewed as well. With the development of marine biology in the 1960s, fisheries management increasingly became an economic issue, that is, concerned with economic performance and not just the size and wellbeing of fish populations. What followed was an important intersection of marine biology and fisheries economics. During the late twentieth century, output quotas (upper limits on the biological outtakes from the ocean) were introduced in many countries as the central management measure, boosting the race for fish:

> Expecting that a quota would be reached during the season and hence the fishery would be shut down, all participants in a fishery have the incentive to catch as much as they can, as fast as they can. This has led to overcapitalization, drastically shortened seasons and losses of life and property. (McCay 1995, p. 4)

This is a phenomenon with many names: derby fisheries, olympic fisheries, capital stuffing or simply the race for fish. In such a situation, vessel owners invest in larger and faster vessels in order to quickly catch a larger share of the limited quota, and the result is overinvestment and so-called overcapitalization. The logical management response to this is to propose some kind of "boat quota" in order to prevent or slow down the race for fish (Moloney and Pearse 1979). In the famous Alaska halibut fishery, for example, the race for fish reduced the 6 months fishing season to less than 48 hours, with fishermen risking their lives in bad weather to get their share of the shared quota. With a boat quota, this is prevented, as each operator is allocated his or her share at the beginning of the season. However, with already overcapitalized sectors and increasing global competition, the boat quota management model had inherited pressure for transferability (McCay 1995). Suffering from bad economic performance, as small total allowable catches were shared between a large number of vessels, operators pushed for a "rationalization" of production—the ability to merge boat quota from two or more vessels. Australia and New Zealand were the first countries to implement transferability of fish quotas in 1983, with Iceland following in 1984 (McCay 1995). In Denmark, an alternative path and distribution method was chosen. Output quotas were introduced in the mid-1970s and early 1980s, and by splitting the annual quota into shorter monthly or fortnightly periods (and later by vessel size and activity), the race-to-fish problem was initially solved. This state-led distribution of fishing opportunities in time-bound "rations" according to size and activity was the basic management principle from the beginning of the 1980s until market-based management was introduced in the period between 2003 and 2007.

A Disciplinary Gap

The implementation of market-based fisheries management in different parts of the world has also increasingly made it the object of cultural and social analysis. The most common themes discussed in this literature are the negative impacts of the concentration of rights, the inequities of distribution, the ethics of privatizing a public resource and, finally, the question of stewardship (Duncan 2011; Olson 2011; Sumaila 2010). The latter debate is centred on the assertion that a sole owner of a resource is a better caretaker; thus, private ownership ensures long-term sustainability (Gordon 1954). This and other questions have led to a prolific academic debate about the social impacts of market-based fisheries management (McCay 1995; Olson 2011), in which there is a considerable disciplinary gap between economists and other social scientists.

Academic critiques of market-based fisheries management have led to an international policy debate, within which it is proposed that market-based systems can be designed to avoid the documented negative social implications in regard to equity and fishing communities (Bonzon et al. 2010; Højrup and Schriewer 2012; Høst 2012; Sanchirico and Kroetz 2010; Schou 2010). This policy design debate is an interesting intersection between non-governmental organizations (NGOs), academics and policy makers. The design debate seems to originate in a dialogue between the promoters and critics of market-based fisheries management. The promoters are most often economists and state bureaucrats, while the opposed criticism originates from NGOs, anthropologists and related branches of the social sciences (Olson 2011; Sumaila 2010; Symes 1996). The economic "pros" and social "cons" of market-based management systems are summed up here by Bonnie McCay:

> Yes, ITQs do result in increased efficiencies, lowering costs of the "race for the fish." Investors can better match capital and labor to the resource itself. On the other hand, the social structure gains new fracture points, co-venturers become owners or labourers, people who thought of themselves as independent fishers begin to use terms like "sharecroppers" and "tenant farmers," or "businessmen" and "fish lords". (McCay in Shotton 2000, p. 42)

Remarkably, fleet concentration and efficiency can be seen both positively and negatively, depending on the priorities of the beholder. The disciplinary gap is therefore more than just a shift in focus, but is instead entrenched in fundamental differences between social science disciplines. Where economists often prioritize individual gains and economic efficiency, anthropologists and ethnologists are concerned with the community and social relations. There is something in the basic assumptions, objectives and instruments of the different sciences that produce and reproduce this gap. An illustration of these differences is the conceptualization of the main resource problem. Put simply, the main issue for biologists is overfishing, which can be avoided by introducing input and output controls. This is where quota management based on virtual population analysis comes in (Holm and Nielsen 2007). However, quota management can bring unpredicted incentives and a race for fish into the system, which in turn may lead to overinvestment and overcapacity. This is an economic problem with resulting bad economic performance, which economists can aim to solve, for example by proposing individual and transferable quotas. The downside of this is the issue of fairness in the industry pointed out by anthropologists, ethnologists and others. Thus, seen from an ethnological point of view, the problem is not overcapacity or overfishing—but overinclusion. There are too many people engaged in the same fishery, and consequently the solution must be to reduce this amount and to find an instrument to exclude people, or alternatively to regulate the size of the amounts allocated to each participant. It is therefore a problem of social organization and cultural values, and anthropologists and ethnologists tend to propose co-management strategies or equality-based rotation systems. The above characterization is of course a simplification, but it still illustrates the different objectives, images and instruments in debates around market-based fisheries management.

The Other Side of Concentration

One of the most controversial aspects of market-based fisheries management is the concentration of fishing rights into fewer hands. Accordingly, this is part of the design of market-based systems, as a means of rewarding the most efficient, profitable fishers,[2] but it also has negative consequences. Fishing activities shift away from self-employed fishers and accumulate to a small number of larger companies. There is not just a quantitative reduction of fishers, but a qualitative change in ownership structures, fishing practices as well as geographic and material manifestations. This will be discussed further in Chap. 6 "Transformation and Modes of Production". Sometimes these quota owners lease quota to active fishers in return for a percentage of their income, prompting the feudal terminology "sea-lords" and "quota barons" (Helgason and Pálsson 1997). Likewise, "slipper skippers" make a living from leasing out quota, while "shore bosses" organize their quota to be fished from other vessels. Measured purely in economic terms, these companies could appear

---

[2] In the Danish case presented in this book it was framed as follows: "The intentions with the arrangement are that quota shares will be concentrated into fewer and more efficient fishing vessels. In this way the fisheries will become more profitable." (Ministry of Food and Agriculture 2010; author's translation)

to be sounder operations with better organization of labour and capital and the potential to compete in an environment of fierce global competition. However, fishing opportunities disappear from fishing communities, as the retiring quota holder did not have to ask the rest of the community for permission to sell his or her quota to the highest bidder. As the number of boats in a community reduces, the operations of others are threatened as shore-based infrastructure becomes unviable. The social landscape changes qualitatively. Young, newly established skippers have to compete on the quota market in order to obtain the fishing rights their fathers' generation is selling. This calls for a closer analysis of social implications when evaluating and recommending market-based fisheries management:

> While many economists have been eager to examine individual transferable quotas, most economic studies of ITQs have focused narrowly on the harvesting process and impacts on boat owners. This has ignored the broader and more important issues of regional and community impacts. (Copes 2004, p. 174)

In Denmark, market-based fisheries management systems have also resulted in a concentration of fishing activity onto fewer vessels. According to the annual report on the economy of the Danish fisheries in 2011 (including the pelagic fisheries), 15% of Danish commercial vessels accounted for over 90% of the total catches measured by volume. Forty-two percent of Danish vessels accounted for more than 90% of total catch measured in value (Fødevareøkonomisk Institut 2011). Out of roughly 700 commercial vessels, 500 with the lowest catch brought in a mere 5% of the total volume of Danish landings. Since 2007, when the market-based system for demersal fisheries was introduced, more than 300 vessels have left the fleet. This reduction in vessels is paralleled with a qualitative change in the social and cultural composition of the fishing sector. What I call *captains of finance* have used market-based management to expand their operations in alliance with legal advisors, accountants, investors and, sometimes, migrant workers. On the other side, there are a large number of share-organized and self-employed fishers who have left the sector or who have adjusted to the fluctuating conditions on the quota-leasing market.

The processes around the annual European Union decisions on Total Allowable Catches (TAC) for each catch area, species and state point to the fact that the balance in the marine ecosystem is now as much a political construction as it is an issue of biology. TACs are subject to political negotiations and lobbyist pressure and can be designed according to political objectives. An assumption underlying this study is that society is also made up of social and economic differences and a plurality of life-modes (Højrup 2003). Which constellation of life-modes and practices should be favoured by the management system, is therefore also a political choice. This book will hopefully prove constructive to this political process by providing an illustration of the social and cultural plurality that is often absent in the economically and biologically influenced perspective in fisheries management.

A Neoliberal Turn?

Privatization and concentration of power have led scholars to interpret market-based fisheries management as part of a wider neoliberal agenda (Mansfield 2004, 2008). Indeed, several characteristics of market-based fisheries management are also central tenets of neoliberalism. A strong emphasis on private rights, free markets and free trade between individuals are all key elements. In fisheries, this extension of free markets and trade not only brings radical new features into the everyday life of fishers, but also to fisheries managers. The role of the state changes from that of the responsible manager and allocator of rights to one of creating and preserving an institutional framework for individual private property rights and trade (Harvey 2005). In neoliberalism, the state as the responsible planner and distributor is partly replaced by the market, which, it is believed, will regulate economic activities in the best way for all (Miller and Rose 2008). A new aspect is thus, that in neoliberalism, new resources and domains of human life are brought into market organization through the creation and use of new technologies. In that respect, TACs and individual quotas are technologies now used to create markets for fishing rights as individual property (Holm 2001). Likewise, fishers in Denmark use the Internet to trade quotas. In neoliberal ideology, if a market for fishing rights does not already exist, it must be created, because it is only through the interactions between individuals in a market, i.e. the boat owners, that the optimum social good can be obtained. By introducing a narrow focus on individual economic behaviour, neoliberal management reforms almost every aspect of the fisheries sector, changing the nature of fishers and communities as well as changing the role of state resource managers.

Neoliberalism of the Oceans

While neoliberal principles in general have gained much attention in academia (Crouch 2011; Harvey 2005; Miller and Rose 2008; Overbeek 1993; Overbeek and Apeldoorn 2012), they have been less studied in academic literature concerned with fisheries (Mansfield 2004). However, in recent years, market-based fisheries management has increasingly been linked to a neoliberal political agenda (St Martin 2007). Since the critical state of world fisheries is as much a crisis of management as it is an environmental crisis, there is a push for institutional change at many levels. In fisheries, the neoliberal answer to this crisis is closely related to the idea of a tragedy of the commons, and market-based fisheries management is a distinct way of answering this problem (Mansfield 2004). It is linked to the belief that individual human behaviour and lack of private property will lead to economic inefficiency and even the destruction of natural resources. Neoliberalism of the oceans is about creating property rights and tying these rights to a market and market rationality (Mansfield 2004). The fishing right as a property reflects an exclusive right to a limited resource. In this regard, it is part of a market-based environmental policy family alongside emissions trading and milk quotas—that is, the right to pollute and produce milk. A distinct feature of neoliberal policies is the attempt to solve the problems caused by a globalized and potentially self-destructive economy by enclosing a limited potential or "good" and by distributing the right to this through market mechanisms.

On the international level, privatization and market-based fisheries management are promoted by powerful organizations such as the *World Bank* and *European Commission on Fisheries*. Often, the Danish VQS system studied in this book is referred to as a best practice example. Similarly, large international NGOs and lobby organizations such as the *Environmental Defense Foundation* and recently the *International Sustainability Unit* as well have used the Danish case in their promotion of market-based fisheries. As argued in a recent report, the two organizations view market-based fisheries management as an investment case:

> Fisheries transition, in many ways, displays similar traits to those of a classic investment turnaround: the upfront costs of transition are offset by the profits that are generated through more efficient and productive fisheries with higher harvests and lower costs. In other words, there is a real return on investment to be had. (Mundy and Band 2014)

By creating an exclusive property, profit can be made from an already overcapitalized fishery. The credibility of these statements in reference to the Danish fisheries are doubtful, though not harmless. As the Danish system is exported and used as inspiration for the rest of the world, it becomes even more crucial to understand it properly and to describe it in a balanced way. This is the task that I set out to accomplish in the present work. It is not the purpose of this book to argue that the Danish VQS system is part of a neoliberal agenda. Rather, I want to examine and understand the processes that take place when a new property right and a new market are introduced. The broader economic context for these processes is a global system in which private companies gain power and nations mobilize their industries and institutions to increase competitiveness (Pedersen 2008). However, for fishers, accountants, quota pool managers and bureaucrats, concerns about market-based policy are all about making it work on a practical level.

In the following chapters I focus on different aspects of the VQS system and market-based fisheries management. Chapter 2, "Growth and Management", is a historical overview of Danish fisheries management, focusing on the growth of commercial fisheries in the twentieth century and the specific problems the Danish government faced when quota management measures were introduced for the first time in the 1970s. This created a distribution problem, and in turn also an overcapacity problem, as the fleet soon had a capacity to fish that was far greater than the available resource. The introduction of the VQS system in 2007 was one way to solve these two problems and the result of a long process with different potential management solutions. The chapter takes a closer look at the elements involved in the growth of the Danish fisheries and their management. More than just technological innovations and open access, growth was fueled by an interplay of state subsidies, tax exemptions, growing markets and competition, a national and international race for fish as well as individual entrepreneurs.

The third chapter "Society and Market" asks the fundamental question, "What is a market economy?" Inspired by the ideas of Karl Polanyi that the market economy was created as a state intervention, which led to the creation of a specific relationship between market and society, I discuss a dialectical relationship between social institutions and social relations on one side, and individual action and gain on the other. From this principal discussion, the chapter moves on to explain the VQS management system as a policy balancing relations between individuals and societal objectives in a market economy. Detailed analysis of the changing industry structure and geographical spread of the Danish cod fishing fleet is presented.

From this general overview of the market, the fourth chapter "The Commodity and its Exchange" shifts focus to the VQS—the quota itself—as a commodity. The chapter examines the VQS as a commodity and looks at the use and exchange-value aspects of the VQS. While the quota share has qualities as a political, administrative and management shaped tool, a look at the use and exchange-value seems productive in order to understand the VQS commodity as a social phenomenon. For example, it is useful to critically examine how vessel owners plan around VQS and its constantly fluctuating value. The chapter goes deeper into the exchange side of the VQS commodity and uses ethnographic material to explain motives behind selling, buying and leasing as well as to question the basis of its value. To understand these complex questions, it is necessary to look at the use-value of quota and different modes of operation.

Thus, the fifth chapter "Access and Fishing Activities" describes and discusses different ways to organize fishing activities. Drawing on ethnographic material, the chapter outlines five different modes of operation, looking at how fishing activities are structured through the year, as a career and the organization on board fishing vessels, all in relation to the VQS commodity. Even though these five ethnographic examples were chosen because of their diversity, the chapter concludes by finding some general patterns related to two principal modes of production, self-employed fishing operations and capitalist organized fisheries. The analysis indicates that the capitalist companies with absentee investors have used market-based fisheries management to expand and invest in fishing quota, while self-employed fishers tend to avoid the financial dependency necessary for expansion.

The sixth chapter "Transformation and Modes of Production" seeks to explain the principal difference between capitalist and self-employed fishers and examines why market-based fisheries management favours large capitalist fishing companies—a group that often perform poorly in equal or open-access management regimes. Here, the concept of a mode of production is used to understand the two different kinds of operations: capitalist organized fisheries and self-employed fishers. Understanding how market-based fisheries management reshapes the fisheries in a more capitalist manner is the key, I argue, to understand the deep transformations that are caused by market-led distribution. From this point of view, the promotion of market-based fisheries is not a promotion of a more efficient fleet, but rather a promotion of a specific set of characteristics, including different life-modes and social relations. Chapter 6 is followed by a postscript "Everyday life and Mediated Fisheries" that reflects on the findings and conclusions of this book in a wider perspective.

In an optimistic manner, the postscript presents a tentative sketch of an alternative management model, framed as a *mediated fishery*—a mix of market mechanisms and state or community control.

References

Bonzon, K., K. McIlwain, C.K. Strauss, and T. Van Leuvan. 2010. Catch share design manual: A guide for managers and fishermen. New York: Environmental Defense Fund.

Brox, Ottar, J. M. Bryden, and Robert Storey. 2006. *The political economy of rural development: Modernisation without centralisation?* Delft: Eburon.

Charles, Anthony T. 2001. *Sustainable fishery systems.* Oxford: Blackwell Science.

Copes, P., A. Charles. 2004. Socioeconomics of individual transferable quotas and community-based fishery management. *Agricultural And Resource Economics Review* 33:171–181.

Crouch, Colin. 2011. *The strange non-death of neoliberalism.* Cambridge: Polity Press.

Duncan, Leith. 2011. The social implications of rights-based fisheries management in New Zealand: for some Hauraki Gulf fishermen and their communities. http://hdl.handle.net/10289/5318. Accessed 16 May 2011.

Fødevareøkonomisk Institut. 2011. *Fiskeriets økonomi (Economic situation of the Danish fishery).* Frederiksberg: Fødevareøkonomisk Institut.

Gordon, Scott H. 1954. The economic theory of a common-property resource: The fishery. *The Journal of Political Economy* 62 (2): 124–142.

Hardin, Garrett J. 1968. The tragedy of the commons. *Science.* Washington, D.C.: AAAS.

Harvey, David. 2005. *A brief history of neoliberalism.* Oxford: Oxford University Press.

Hasslöf, Olof. 1949. *Svenska västkustfiskarna. Studier i en yrkesgrupps näringsliv och sociala kultur.* Göteborg: Svenska västkustfiskarnas centralförbund.

Helgason, Agnar, and Gísli Pálsson. 1997. Contested commodities: The moral landscape of modernist regimes. *Journal of the Royal Anthropological Institute: incorporating "Man"* 3:451–471.

Holm, Petter. 2001. *The invisible revolution: the construction of institutional change in the fisheries.* Tromsø: Norwegian College of Fishery Science, University of Tromsø.

Holm, Petter, and Kåre Nolde Nielsen. 2007. Framing fish, making markets: The construction of individual transferable quotas (ITQs). *The Sociological Review* 55:173–195. doi:10.1111/j.1467-954X.2007.00735.x.

Højrup, Thomas. 2003. *State, culture and life-modes: The foundations of life-mode analysis.* Aldershot: Ashgate.

Højrup, Thomas, and Klaus Schriewer. 2012. *European fisheries at a tipping point* Estudios Europeos, No 1. Murcia: edit.um.

Høst, Jeppe. 2012. Fairness or efficiency? *SAMUDRA.* 61:15–17.

Mansfield, Becky. 2004. Neoliberalism in the oceans: "Rationalization," property rights, and the commons question. *Geoforum* 35 (3): 313–326.

Mansfield, Becky. 2008. *Privatization: Property and the remaking of nature-society relations.* Antipode book series. Malden: Blackwell Publishing.

McCay, B. J. 1995. Social and ecological implications of ITQs: An overview. *Ocean and Coastal Management* 28 (1): 3–22.

McCay, B. 2001. Environmental Antropology at Sea. In *New directions in anthropology and environment: Intersections,* ed. Carole L. Crumley. Walnut Creek: AltaMira Press.

Miller, Peter, and Nikolas S. Rose. 2008. *Governing the present: Administering economic, social and personal life.* Cambridge: Polity Press.

Moloney, David G., and Peter H. Pearse. 1979. Quantitative rights as an instrument for regulating commercial fisheries. *Journal of the Fisheries Research Board of Canada* 36 (7): 859–866.

Moustgaard, Poul H., and Ellen Damgaard. 1974. *Garnfiskere: Organisation og teknologi i et vestjysk konsumfiskeri*. Esbjerg: Fiskeri- og Søfartsmuseet, Saltvandsakvariet.

Mundy, Justin, and Laurence Band. 2014. Towards investment in sustainable fisheries: A framework for financing the transition. New York: Environmental Defense Foundation & International Sustainability Unit.

Olson, Julia. 2011. Understanding and contextualizing social impacts from the privatization of fisheries: An overview. *Ocean & Coastal Management* 54 (5): 353–363.

Ostrom, Elinor. 2002. *The drama of the commons*. Washington, D.C.: National Academy Press.

Overbeek, Henk. 1993. *Restructuring hegemony in the global political economy*. London: Routledge.

Overbeek, Henk, and Bastiaan van Apeldoorn. 2012. *Neoliberalism in crisis*. Houndsmills: Palgrave Macmillan.

Pedersen, Ove K. 2008. *Institutional competitiveness—How nations came to Compete*. In *Working paper*, ed. International Center for Business and Politics. København: Copenhagen Business School.

Rasmussen, Holger. 1968. Limfjordsfiskeriet før 1825: sædvane og centraldirigering. København.

Sanchirico, James N., and Kailin Kroetz. 2010. Economic insights into the costs of design restrictions in ITQ programs Washington: Resources for the future.

Schou, Mogens. 2010. Sharing the wealth. *SAMUDRA* 55:18–23.

Shotton, R. 2000. Use of property rights in fisheries management: Proceedings of the FishRights99 conference, Fremantle, Western Australia, 11–19 November 1999. Rome: Food and Agriculture Organization of the United Nations.

St Martin, Kevin. 2007. The difference that class makes: Neoliberalization and non-capitalism in the fishing industry of New England. *Antipode* 39 (3): 527–549.

Sumaila, Rashid. 2010. A cautionary note on individual transferable quotas. *Ecology and Society* 15 (3): 36.

Symes, David. 1996. Fisheries management and the social sciences: A way forward? *Sociologia Ruralis* 36 (2): 146–151. doi:10.1111/j.1467-9523.1996.tb00011.x.

Chapter 2
Growth and Management

Abstract This chapter seeks to reframe the reductionist historical narrative explaining problems of overfishing with a combination of human economic behavior, technological innovations, and lack of property rights. Instead, contemporary Danish commercial fisheries are seen as part of a more complex long-term development, with close attention paid to the actions of state, management, organizational politics, private companies, fishers, and their communities. The aim is to show the interplay of a broad range of factors, structures, and actors that influence fisheries management. Thus, the chapter seeks to avoid reducing the history of quota privatization to a simple and deterministic narrative of technological development and the tragedy of open access. The history of fishing is much more nuanced than this. This inquiry therefore looks more closely at, behind, and around the axioms of fisheries management narratives.

Keywords Fisheries management · Technological development · Share organization · Individual transferable quotas · Overcapacity

Long Past, Short History

It often appears as if fishing has a long past but a short history. While fishing is one of the oldest occupations in the world, it has undergone immense development in the last 100 years or so. Traces of its past go back many thousands of years to the times before human societies were changed by agriculture and urbanization. However, when the histories of today's issues in fisheries management are written, it is not this long past that is depicted. Rather, what we encounter not only in the media but often also increasingly in academia is a much shorter history, a story of growth in fishing capacity and decline of fish stocks, a story of constant innovations in technology and open access tragedies. When writing about the history of the modern fishing fleet, and in particular the problems of present-day fishing, the narrative tends to begin at around the end of the nineteenth century. Since that time, technological inventions such as motorization, decked boats, new nylon-based materials, advanced navigation, and fish finding equipment have changed the fishing sector many times over and created the modern efficient fleet (Fig. 2.1). Throughout the

Fig. 2.1 The immense growth in tonnage, new technologies, and lack of proper management are the central tenets in the common narrative. In this deterministic explanation, the resource depletion in the twentieth century is caused by individualistic human behavior, technological innovations, and lack of property rights. Steel trawlers like these, here anchored to massive concrete harbors, are on the other side perceived as the natural response to increased global competition. (Photo: Jeppe Høst)

twentieth century, the Western fishing fleet has experienced an unceasing technological development and growth in size, fishing effort, and capacity. According to the dominant historical narrative, during this period the combination of open access to fisheries, technological growth, and human rational behavior led to an unavoidable resource collapse. Today, this has fostered widespread international concern about the health of ocean ecosystems. In the last decade, several studies have been published showing that a majority of the world's fish stocks are fully exploited or overexploited (Zeller and Pauly 2007). These go hand in hand with media and newspaper headlines such as "Last fish to be caught in 2048" (a 2006 article in *Science*) and "Just 100 cod left in North Sea" (in *The Telegraph*, 16th September 2012). This narrative of "boom and bust" fisheries is deeply entrenched in academia and across the media, among NGOs and in the public sphere. Indeed, there may be a self-perpetuating synergy in the relationship between the dramatized historical narrative and the media, NGOs, science, funding organizations, and governments. But if the history of present-day fishing can be told as a narrative of rapid technological growth in the twentieth century and the tragedy of open access, then the theme of this book, market-based fisheries management, is introduced as the logical solution to this historical development. In this dominant narrative the introduction of individual property rights to our limited maritime resources has become the inescapable solution to the persistent growth in fishing capacity and to the open access nature of wild capture fisheries.

On the other hand, it can convincingly be argued that advanced, large-scale fishing has existed for centuries. The Basque distant-water cod fishing off the coast of Newfoundland in the sixteenth century and the large-scale Dutch herring fisheries in the North Sea in the fifteenth century are just two of the most prominent examples. Both of these fisheries were based on new discoveries in technology, and an advanced organization was necessary to carry out their complex, geographically diffuse activities. Tracing the beginning of the history of modern fishing to the late nineteenth century therefore excludes these and many other fisheries and technological developments. Worse, it also hides an ongoing and dynamic relationship between different types of fishing fleets that compete and collaborate with one another and employ a wide range of gear types and fishing methods in order to sustain livelihoods in an ever-changing social and natural environment. In the wider perspective, the history of fishing is a story of environmental and seasonal fluctuations leading to shifting possibilities and relative advantages between operators and regions. Rather than one "boom and bust" narrative, a closer historical examination reveals several cycles of boom and bust. The Danish fisheries are considered to have peaked in the 1500s and then to have declined before rising to prominence again in the late nineteenth century (Vaarning 1984). Fears about the health of fish populations due to declining catches were recorded in the 1600s and again in the late nineteenth century, when fish stocks were reported to have disappeared from near-shore waters (Mortensøn 2004). In the largely enclosed fjord Limfjorden, the familiar story of growth and resource decline was played out, but the action took place at least 100–200 years earlier than our dominant narrative of twentieth century growth. In the Limfjorden the human consequences of the crisis in fish stocks led to debates at the state level about effective resource management, disputes over property and specific types of fishing gear, and methods being accused of doing damage to the natural environment, all topics familiar today (Rasmussen 1968; Stoklund 2000; Østergaard 1984). In 1741, the first of several commissions was appointed to undertake a study of the wellbeing of juvenile fish in *Limfjorden*. The commission pointed to the negative effects of purse seine nets and created closed areas that remained in place until around 1900. The exploitation of fisheries in Denmark thus has a long history, and there is vast historical evidence of growth and management preceding the twentieth century. It is only by closely examining this historical process that we can fully understand the full range of developments that led to the introduction of a market-based fisheries management system at the beginning of the twenty-first century.

The Vessel Quota Share (VQS) system has, in many ways, shed new light on the history of Danish commercial fishing by highlighting the need to study the role of management (and in particular quota management) as an active determinant of fishing practices. Former sociological and anthropological studies of Danish fisheries have largely focused on a specific fishing community, on the role of technology or the local consequences of the resource crisis and its management.[1] Rarely does

---

[1] Most of the community studies into Danish fisheries were conducted in the 1970s and 1980s before output and quota management were introduced. The management and regulation of fishing is seldom mentioned. The study of technological developments often illustrates the important role of

fishery management have a central position in the research, and rarely is it treated as much more than an external and necessary condition for fishers and fishing communities. But as quota shares are now transferable commodities, and as they have fundamentally changed the nature of access to fish resources, the specific history of fishing rights and their role in the management system have become central to social-scientific understanding of fisheries. To understand the creation of the VQS system, one unavoidably also has to understand environmental regulations, fishing-effort management, and the internal dynamics of the fleet. The VQS system thus allows new perspectives on that history. To reframe and understand the growth and management of Danish fisheries in its complexity is therefore the aim of this chapter. Finally, this chapter is also a chance to write about the history of the Danish commercial fishery for an English speaking audience. This contributes to wider international debates about the history of modern commercial fisheries, which serve to illuminate and question the historical terms through which market-based fisheries management is often promoted. The VQS system has radically altered the premise upon which the modern development of commercial fishing in Denmark is based, and quota shares, quota markets, and quota leasing will take up a natural and perhaps more prominent part in any future descriptions of contemporary Danish commercial fisheries.[2]

Sharing

Share organization is a characteristic of Danish fisheries that reaches back into the distant past. The share principle is still used today to such a degree that it has been mainstreamed in Denmark as the legal structure of fishing labor organizations. The share principle means that the crew members onboard a vessel are paid in shares (a certain percentage) of the income earned on each fishing trip. With the uncertainties of nature and the shifting of the seasons, sharing is a suitable way to organize fishing activities, and its use is in no way limited to Denmark (Højrup 2002; Löfgren 1977; McCay and Acheson 1987). The notion of sharing has historically been central to Danish fishing operations. Not only as the sharing of the catch or income but also as the sharing of all aspects of production. In 1880 a public authority pub-

---

state loans and subsidies but not the management aspect (Hjorth Rasmussen 1984). One exception is the PhD thesis of Morten Karnøe Søndergaard, which includes a study of technology within a management, organizational, and political contextualization (Søndergaard 2004, 2008).

[2] Earlier studies often focused on a specific community, and unfortunately the management and regulations of access were often absent or left somewhat untouched. Some publications give a national and historical overview, but they often concentrate on the technological and quantitative developments. Recently a group of museums in coastal regions have published documentation on the development of the fisheries in a number of communities (Holm 1994, Byskov 2010). Compared to other nations, such as for instance Norway or Iceland where fishing represents a larger part of the economy and national identity, Denmark has little coherent literature on the development of the fishing sector and the management dimension in particular.

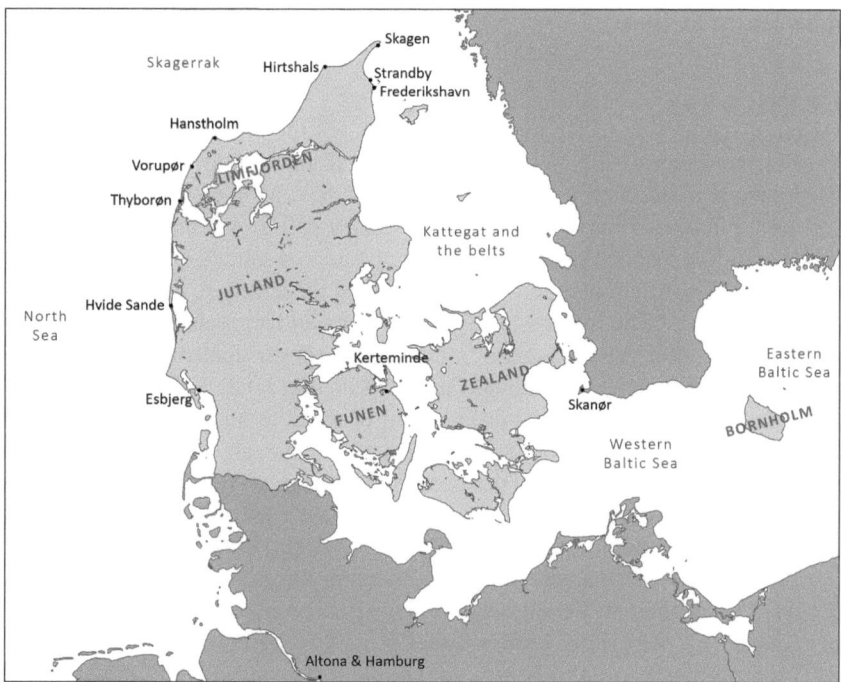

Fig. 2.2 Map showing Danish fishing towns and the fishing areas around Denmark

lished a report on the organization of fisheries in the remote community of Vorupør in Western Jutland (see Fig. 2.2; Hjorth Rasmussen 2000). There were no decked boats, only 14 open boats, each with six crew members. When fishing, each of these men had to meet in the morning with four sets of baited hooks. Each set consisted of a longline with 60 snoods and hooks. These hooks were set close to the coast and outwards. At the end of the fishing trip the catch was shared equally and distributed individually, often by the fishers' wives or children.

The principle of sharing is easily identifiable here. Not only did the six men share a boat but they also shared the burden of baiting and preparing fishing gear. At the end of the trip, they even shared the catch and the tasks of distributing and selling it. Both in relation to the baiting process and the distribution of the catch, the wider family was an integral part of production. The boat was shared as the production unit, as was the reproduction process of maintaining and preparing gear. By adopting the share organization, fishers also shared the risk of an unsuccessful fishing trip as well as the potential benefits of a successful fishing season and good sales. It is from share organizations like this that present-day share-based fisheries have inherited their organizational model. Even though the large investments in technology and developments in gear have led to changes in the traditional form of share organization, the sharing principle is still identifiable. The economic and material burdens of the vessel, fishing gear, and bait are shared, as well as the outcome of a fishing trip, whether successful or not.

Technology at Sea

In the short history of fishing, motorization plays a key role as the initiator and basis of intensive growth of the industry. In Denmark the motorization of the fishing fleet occurred around 1900. Within a few years, from 1900 to 1905, the total catch in Denmark was doubled to 70,000 t (Holm 2002). Motorization brought many advances for fishers after centuries of moving by sail power and rowing. Fishing grounds could be reached faster and the catch could be landed quicker, resulting in a better price. The motor also increased safety and the ability to navigate in difficult weather conditions. The result was higher mobility and a greater operational range. However, if we look more closely at motorization (or mechanization, as it can also be termed), the first engines installed onboard fishing boats in Denmark were not actually used for transportation purposes but in order to haul in fishing gear. This was particularly the case in the development of the Danish seine fisheries,[3] where the seine had to be hauled in with the catch inside the net (Hjorth Rasmussen 1984). This hauling motor helped lift the weight of the catch, and it was only later linked to an outboard propeller through a "cycle-chain" system. Based on experiences with this setup, sternpost propellers (with a connection through the ship to an onboard engine) were installed on fishing vessels. These early motors were steam powered, later replaced by petroleum, before diesel proved to be the most reliable fuel. In 1889, for example, a group of fishers applied for a state subsidy to fund a motor for their new boat. According to the fishers, they needed a powerful engine to drag in their Danish seine, as the competition forced them to reorganize their fishery from close coastal operation into a longer-range seagoing fishery (Hjorth Rasmussen 1984). Working with larger gear at greater depth meant a heavier workload. What we can learn from this example is that already in 1889 growth, competition and state subsidies played a part in the development of fisheries—even before engines became commonplace on board fishing boats. In fact, since the 1860s the larger and more expensive deck boats had been developed with state subsidies and sometimes even state ownership (Hjorth Rasmussen 1984).

One of the catalysts of the Danish expansion in the second half of the nineteenth century was the development of the Danish seine. This expansion has to be understood in the context of the absence of a proper Danish fishery in the North Sea. This created the impetus for the Danish state in granting subsidies for experimentation and new initiatives. As state investment in North Sea fisheries grew, fishers in other parts of the country—such as the fishers from the eastern region of Bornholm—were increasingly being declined in their applications for state grants. These were dedicated instead to developing the seagoing fishery that the ministry considered to be in the strategically important North Sea (Hjorth Rasmussen 1993). In the North Sea, British and German fishers were often seen close to Danish coasts, while the coast of Western Denmark (Jutland) was considered to be "undeveloped country" in regards to fishing (Meesenburg and Højrup 1984). During the nineteenth and

---

[3] Danish seine, sometimes also called anchor seining, is a seine haul technique based on the beach seine. Its development in 1848 is credited to the Limfjord fisher Jens Væver.

twentieth centuries the fishing activities in the North Sea and Skagerrak, previously seen as too minor to be included in statistics, grew to become the economic backbone of the Danish fishery (Hjorth Rasmussen 1984). Landings from the North Sea increased from 10% of the total Danish landings in 1900 to 70% in the 1960s. This was paralleled by a continuous decline in catches from internal waters (Kattegat and the belts) and the stagnation of fishing activities in the Baltic Sea.

This shift would have been impossible without the early application of state subsidies. While fishing companies from both Germany and England had large operations fishing in the North Sea, often seen from the coast, Danish large-scale fishing operations were rare until later in the twentieth century. Some companies tried to attract greater capital to Danish fishing activities, one example being "Esbjerg Aktiefiskeriselskab," a stock company that was established in Esbjerg in 1879. The company had three deck boats for line fishing and one steam powered trawler, but economically the company was a failure and was dissolved less than 4 years later (Hjorth Rasmussen 1984). Through the subsidies directed to smaller producers, the state therefore played a crucial role in the shaping and expansion of Danish fisheries, even before motorization. The Danish seine technology—developed at first single handed by the fisher Jens Væver—was an important step in this expansion of Danish fishing effort into the North Sea. Seine nets replaced the expensive, large and difficult trawl method with an inexpensive and lighter technology that was much easier to handle. Small share-based units could now take part in the fisheries that were previously limited only to companies with the capacity to invest in expensive trawling gear. What is peculiar is that this development was shaped through small loans and grants for the testing of new boats and motors, and not through large-scale funding of operators or scientific ventures. State subsidies were part of an organizational setup with local fishermen's organizations and what we can call *captains of industry*—the inventive entrepreneurial people closely engaged in both the actual operation of fishing vessels and the development of inventive new fishing techniques (Veblen 1964).[4] The introduction of the Danish seine gave fishing communities the possibility to increase their fishing and income and provided the basis for the development and use of deck boats,[5] initiating growth and expansion that preceded that of motorization. The increases in catches connected to the mechaniza-

---

[4] The captain of industry is in Thorstein Veblen's account opposed to the absentee owner, where the managerial tasks take the "captain" away from the actual process of production. It should be noted that Veblen argues that this characterization is a myth and is actually based on far fewer people than the popularity of the characterization suggests. "In the beginning the captain was an adventurer in industrial enterprise—hence the name given to him; very much as the itinerant merchant of the days of petty trade had once been an adventurer in commerce. He was a person of insight—perhaps chiefly industrial insight—and of initiative and energy, who was able to see something of the industrial reach and drive of that new mechanical technology that was finding its way into the industries, and who went about to contrive ways and means of turning these technological resources to new uses and a larger efficiency; always with a view to his own gain from turning out a more serviceable product with greater expedition." (Veblen 1964, p. 102).

[5] Deck boats were known in other regions and countries, but the development of a deck boat suitable for landing and hauling onto the beach was conducted by a ship builder in Vorupør around 1889–1890 (Hjorth Rasmussen 1984).

tion and motorization of the fisheries in the period from 1880 to 1910 can be termed the first industrial revolution of the fisheries (Søndergaard 2004). A second revolution took place after the Second World War and involved immense developments in materials, fish finding equipment, and the use of hydraulic power. These will be touched upon later. The great plaice fisheries in the 1880s and onwards, which were based on the new use of Danish seine in Skagerrak and in the North Sea, renewed a debate about the need to construct new harbors on the windy and sandy west coast of Jutland. In this aspect, state subsidies and state decisions were also crucial in determining the shape of Danish commercial fisheries and regional economies.

Harbors and Land-Based Infrastructure

Motorization and mechanization, combined with the spread of the Danish seine technique, created the potential for growth in the Danish commercial fleet. However, if the fishing fleet was to develop into a proper seagoing fishery and take part in the highly productive fisheries of the North Sea and Skagerrak, then Northern and Western Jutland would need proper harbors (Rasmussen and Hjorth Rasmussen 1972). As long as the vessels were landed on beaches by human power, fishing activities would not be able to expand much beyond that of small deck boats. At the same time as new harbor building techniques made these developments possible, the expansion of the railways made it reasonable to construct harbors on the remote west coast. Increasing amounts of fish could not only be landed but also transported to the important urban markets in Denmark as well as in northern Germany. The patterns of these harbor constructions were shaped by political decisions and local lobbyism, and their development adhered to a range of constantly changing conditions in the whole maritime sector as well as to infrastructural improvements on land.

For example, at Hirtshals a breakwater pier structure was constructed in 1879 to give lee to approaching fishing vessels landing on the beach. It was lengthened shortly afterwards, and by 1900 it extended out 276 m. At that time, there was only one larger vessel fishing from Hirtshals, a 50 t boat owned by a merchant from neighboring Frederikshavn. The remaining fleet consisted of smaller boats of around 5 t (Vandsted 2004). A few years later, in 1908, local fishers called for a real harbor in order for the fishery to develop and be competitive (Vandsted 2004). This development of the fishery required larger vessels, and for that they needed a proper harbor. After political discussion the construction of a new harbor commenced in 1920 and lasted the remaining decade. However, the harbor had hardly been completed in 1930 before a resolution from the fishers demanded an even larger harbor (Vandsted 2004). The railroad connection in 1925 had improved the distribution channels and increased the price of fish at the auctions. Since then, the harbor has been expanded step-by-step, most notably in the 1960s. The sale of fish for domestic consumption still plays an important role in the harbor's activities, but this is now supplemented with offshore oil and gas industries, shipping and the industrial reduction fisheries, the latter which will be discussed at greater length below.

In the 1960s, Faroese and Icelandic purse seiners began landing huge amounts of herring and mackerel in Hirtshals, and the town consequently built up an industry increasingly dependent on foreign landings. This had a negative impact on prices for many local operators, but it benefitted the local fish processing industry and the many people working there. As vessels grew larger in order to remain competitive, there was a need for a larger and deeper harbor. In 1960, Hirtshals harbor was approximately 4.7 m deep. Today, the largest fishing vessels require 9 m of depth. The Hirtshals example shows the complex set of relations that go into the making of a harbor. Local needs merge with demand from international vessels and the dynamics set by a global market. Other examples, especially those from Western Jutland, show similar trends, all with their local peculiarities. Hanstholm, a harbor that now aims to be the "fishing port of Europe," was recognized by law in 1917, but only in 1967 did a proper harbor replace the breakwater structure built in 1911. In the following decades the harbor was expanded incrementally (1977, 1987, 1997, and 2006), and the seafood, ferry and freight industries were established next to the first harbor basin. Esbjerg harbor came into legal being in 1868 as a replacement for the export harbor in Altona lost in the Second Schleswig War of 1864. Esbjerg harbor quickly grew into one of the largest fishing harbors in the world and had a major impact on local demographic shifts, attracting people from the inland and coastal areas north of Esbjerg. In 1900 more than half of the fishers in Esbjerg were from Holmslandklit and beyond (areas north of Esbjerg). This demographic trend continued until harbors were constructed in those areas suffering out-migration, most importantly in Hvide Sande in 1931 (but also Thorsminde in 1932 and Thyborøn in 1918).

With the new harbors and the expanding railroads and road system, many places grew from mere *places to fish from* into permanent settlements of established fisher communities (Moustgaard and Damgaard 1974). Harbors and fishing activities thus had an impact on everyday life in the remaining towns and uplands. The Danish seine fisheries transformed places like Frederikshavn, and fishers, previously among the poorest members of society, became an esteemed class (Hjorth Rasmussen 1984). The increased incomes from fishing were visible in better houses and clothing. This would not have been possible without the rising demand for fish in inland and urban areas in Denmark coupled with the introduction of the infrastructure necessary to bring this perishable product to consumers.

Markets, Demand, and Distribution

The technological progress in sea and land-based infrastructure would be of no economic use without consumers or demand for fish products. Preservation techniques and transportation networks played a crucial role in developing the demand side of fish markets. Railways are often considered a catalyst for local fisher communities in Denmark, providing access to much greater markets than existed locally. From the second half of the nineteenth century, railroads increasingly provided a link between urban markets and what had previously been remote fishing communities. Growth in the fishing sector accordingly had its parallel—and precondition—in urbanization and in rising mass consumption (Wilcox 2006). Demand and distribution were central to fishing activities, since without a buyer the increasing volumes of fish caught would be of almost no monetary value. For many remote and isolated fishing communities, local demand had previously been an absolute limit on both fishing activities and income. These remote communities also experienced rise and falls in demand and prices, prior to the establishment of the railways. Before tar roads and railroads, it was often wives or children who went out on foot from farm to farm to sell the catch, which often meant that prices quickly fell in favor of the buyer (Hjorth Rasmussen 2000). Given the right conditions, the catch could be sent by ship, in so-called "well smacks," to the nearest large town (Hjorth Rasmussen 1968). Another option was to transport the catch alive in a dam on board the fishing vessel. This was only practical for a limited number of species, most importantly the flatfish species like plaice, but also cod and eel. Seaborne transportation limited the distribution to the coastal markets with their own local suppliers, and this was precisely the obstacle that railroads could overcome. For example, fish loaded onto the train in Frederikshavn in the morning could reach the market in Hamburg the next day at noon. However, even with these new distribution options, fishers sometimes opted to collectively limit their landings in order to achieve a better price for all (Hjorth Rasmussen 1984; Vandsted 2004).

The new distribution opportunities also created new demand for different fish species. In Hvide Sande, for example, new access to German markets was the basis for a sole fishery (Moustgaard and Damgaard 1974). As railroads and road systems allowed products to reach more distant locales, new processing, packing, and preservation techniques also created new uses for the produce of fishers. A local fish meal factory could greatly increase demand in a local area, as could the invention or introduction of a new preservation technique, such as canning. The appropriate preservation of the catch would mean that even without railroads, fish could reach a much wider market. For example, prior to the finalization of the local railroad in 1900, the town of Kerteminde had several different distribution channels. In the 1840s, smoked mackerel was exported to Hamburg, and a new salting facility had improved prices and stability of demand. In 1862 a fish guano (fertilizer) factory was established, and in 1890 a cannery was built. Only in 1900 did the railway become a determining factor for distribution (Vaarning 1984).

For many fishers, the increased access to inland and urban markets meant more than just a larger turnover; it was the factor that enabled fishing to become a full-time occupation (Holm 2002; Tophøj 1976). Prior to the end of the nineteenth century the authorities in Kerteminde did not recognize fishing as a legitimate occupation, even though most of the town was somehow involved in it and fed by it (Vaarning 1984). With the urbanization of the nineteenth century, fishing as a profession was made possible, and with that also came the professionalization of fish processing. During this period, herring catches from Kerteminde were sold to places as far afield as the Caribbean islands (Vaarning 1984).

Globalization and Competition

Railroads meant that harbors were connected to inland and urban markets, allowing for much faster transport of the perishable fish; in many places this is now done via roads and trucks. Being connected to a larger market meant a significant rise in the quantities that fishers could catch and sell without depressing prices. But this linking of remote coastal communities to larger urban areas also had a downside. Suddenly, external competition found its way into remote communities. Investments in vessels and gear had to be paid off with income from prices set by the most competitive operators in the region. This was the double-edged impact of economic globalization. Advances in transportation meant both access to a greater market and an increase in competition (Højrup 1984). With the globalization of markets and distribution systems the competition in local waters became more and more shaped by global dynamics. For example, already in the 1960s the prices paid for plaice caught in Denmark were influenced by those of fish products coming from South America; and today the most direct price competitor in supermarkets is Vietnamese fish. Another impact of the globalization process was a shift in consumer habits and eating patterns.[6]

The result of this economic globalization was a market-driven imperative for the costs of production to be reduced. This was achieved by increasing the average size of fishing operations, introducing new technologies, larger gear, expanding a vessel's operating range or targeting new species. This process, a constant search for comparative advantages in relation to the rest of the fleet, both locally and globally, led to the expansion of European fisheries into more distant waters, to greater depths, and in the number of species targeted. Eventually this developed into an intense race for fish in international waters, resulting in the establishment of exclusive economic zones (EEZ) in the 1970s.

Cannons and Commons

Disputes over the ownership of maritime resources and territories were not new in the twentieth century. For the large medieval fisheries such as Skanør, often from a single locality, sea-borne trade was an integral part of operation. More than just the vicinity to lucrative fishing grounds, these fisheries were dependent on shifting trends in maritime navigation, international relations, and trade. Therefore the

---

[6] Another aspect of globalization is the spread of eating habits. These habits influence not only how fish are processed and prepared but can also greatly increase the value of fish species. In Danish waters, for example, bluefin tuna was historically considered a nuisance. Mainly caught as a sports fish for leisure, its only market was as a trash fish for pig feed. Only during the world wars could tuna be sold for human consumption. Bluefin tuna disappeared from Danish waters in the 1960s before the introduction of quotas and before sushi became a popular meal choice in Denmark.

ability to navigate freely and the ownership of maritime resources were central issues for the fishing activities of the past, just as they are now. From the 1500s, the European seas were subject to constant struggles over sovereignty and conflict over open and closed seas, with shifting nations controlling ocean territories. The Spanish and Portuguese domination of the seas disintegrated in the seventeenth century with the economic growth in Great Britain and the Netherlands and their role in international seaborne trade (McCay and Acheson 1987). During this period, the Mare clausum doctrine was slowly substituted with Mare liberum,[7] but with the important exemption of a national three-mile exclusive zone. The three-mile zone, the political compromise between "requirements of national security with freedom of trade and navigation" (Nandan 1987),[8] was based on the range cannons could effectively reach. From the beginning of the eighteenth century these three-mile zones were agreed on and adopted by most maritime nations, and their remit applied to more than national security. They soon defined a narrow strip of coastal waters within which (among other things) nations had the exclusive right to protect fishing resources from foreign exploitation.

Expanding EEZs

The fishery-dependent nation Iceland is well known for multiple expansions of its EEZ since the 1950s and the dramatic Cod Wars with Great Britain that followed these unilateral decisions. However, outside Europe, expansions of the marine territories of the USA, Peru and Chile preceded those of Iceland. Through the Truman Proclamation of 1945, the USA declared the right to protect and conserve current and future fishing interests *in those areas of the high seas adjacent to the coasts of the USA*.[9] In 1947, Peru and Chile were the first states to proclaim an economic exclusive zone reaching out 200 miles from the coast. In Europe similar discussions were ongoing at this time as to the size and configuration of national EEZs. For Europeans, EEZs were considered to be mainly a fisheries issue, and neither the 1958 nor the 1960 Geneva Conventions managed to solve the issue (Nandan 1987). Only with the 1964 European Fisheries Convention was a substitute to the three-mile zone adapted. The convention established the rule that each coastal state had

---

[7] The term mare clausum is used to mention a sea, ocean, or other navigable body of water controlled by a state that is closed or not accessible to other states. In contrast, mare liberum is a term for a sea that is open to navigation to ships of all nations.

[8] http://www.fao.org/docrep/s5280T/s5280t0p.htm (Accessed August 12, 2012).

[9] The Second Truman Proclamation from 1945: "In view of the pressing need for conservation and protection of fishery resources, the Government of the United States regards it as proper to establish conservation zones in those areas of the high seas contiguous to the coasts of the United States wherein fishing activities have been or in the future may be developed and maintained on a substantial scale. Where such activities have been or shall hereafter be developed and maintained by its nationals alone, the United States regards it as proper to establish explicitly bounded conservation zones in which fishing activities shall be subject to the regulation and control of the United States."

exclusive rights inside a six-mile belt; while in the zone between 6 and 12 miles, other states with a history of fishing in the area (based on activities between 1953 and 1962) had the right to continue doing so. This was probably the first use of the principle of historic catches as a pragmatic tool for defining fishery management in a European political context.

At the same time, Iceland was gradually extending its EEZ, though not without conflict and resistance from Great Britain. In four moves from 1950 to 1975, Iceland unilaterally expanded its EEZ from 3 to 4 nautical miles[10] to 12, then 50, and finally 200, following the standard set by South and North American states. The Icelandic economy relied a great deal then (as now) on incomes from fishing, and perhaps its geographic isolation and distance from other fishing nations made the unilateral moves practicable. The actions, however, led to clashes with Great Britain, which only accepted the last expansion after international pressure (following Iceland's threat to close down an important North Atlantic Treaty Organization (NATO) base on the island). This was a period when fishing and international politics were closely tied (see also Karlsdóttir 2005; Rozwadowski 2002; Finley 2011; Finley 2007).

In 1982, most countries in the North Atlantic region had signed the UN Convention on the Law of the Sea (UNCLOS) that mainstreamed the 200 nautical miles EEZ. This led to detailed negotiations and discussions on the exact shape of the midlines between European countries. The weight of small remote islands in relation to the location of midlines was suddenly vital, and the bilateral discussions between countries involved strategic alliances and deals (Bøggild 1983; Cardwell 2012). The development of the EEZs had a concrete impact on fisheries management in Northern Europe. When the radical expansion of EEZs through the 1970s became more and more politically realistic, fisheries managers began to prepare and look for alternative solutions. The expansion of EEZs would radically change the mobility and seasonal dynamics of the fishing fleets. Nowhere was this as acute as in the North Sea; surrounded by several strong fishing nations such as Denmark England, Scotland, Norway, France, Germany, and the Netherlands—and with visiting fleets from other countries—the entire North Sea would be split into exclusive zones hindering the mobility and activities of existing fleets. Neither fishing practices nor the movements of fish stocks were aligned with the new borders. Thus the coming expansion of EEZs gave a push for the introduction of national quotas. With a system of total allowable catches (TACs) decided on the international level and later split into national quotas based on the historical catches (today known as the relative stability), it was possible to bypass the negative effects of the 200 miles EEZs. Each nation could carry on with its share of the TAC, and national and local peculiarities could be maintained even though they took place in another nation's EEZ. TACs and national quotas, implemented for the first time in 1974 for the herring fishery, allowed both international management and national sovereignty. From the first introduction in 1974, the quota principle quickly spread to other species and fishing areas in and around Denmark as well as Northern Europe.

---

[10] Because the baseline upon which the 4-mile zone was mapped out was also simultaneously changed, the consequences of the extension from 3 to 4 miles were graver than indicated in the 1 mile change.

The period from the 1960s to the 1980s was an era that reshaped the geopolitical seascape, laying the necessary foundations for later fisheries reforms. As a part of this, the vast majority of high-seas fish stocks came under the control of a multitude of countries, and they were no longer subject to free utilization by the most industrial (or industrious) seafaring nations. Market-based fisheries systems would be worth little without substantial national sovereignty over fisheries resources. With national control in place and the recognition of historic catches as a distributive principle, some of the preconditions were in place for the commoditization of fisheries.

Even though this period established the conditions for market-based fisheries management systems, this process did not silently follow a predefined route towards individual transferable quotas (ITQs). Debates took place in the European scientific milieu during this period on whether to choose TACs or *catch per unit effort* as the basis for biological modeling. According to some, the TAC regime was chosen not because of its scientific superiority but because it could produce fair and stable solutions in the international political context, and because it assigned defined roles and responsibility to researchers and managers (Hersoug 2005; Holm 2001). It is interesting to note that in the division between international and national management the quota instrument proved to have similar qualities.

The TAC approach and quota principle therefore are more than mere management instruments—they are also linked to politics, geography, and technology. The combination of EEZs, TACs, and national quotas made it possible to exchange and swap quotas, therein allowing the first degree of commoditization. On behalf of Danish fishers, the Danish state exchanged fish quotas with neighboring nations. Cod quota in one area could be exchanged with sole quota in another; and through this swapping of fishing quotas, managers could find some flexibility and extra resources for the growing sector. The international nature of the affair and the European race to fish had consequences, even at the local level. In local organizations the topics raised at board meetings were increasingly concerned with international issues and fisheries politics at the EU level. It was no longer possible for fishers to be self-managing *captains of industry*. They were met with more and more restrictions at the fishing grounds, restrictions that were counter to the previously taken-for-granted principle of free access to fish (Vandsted 2004). What was taking place was a closure of the seas, not only through EEZs but also through quota management restrictions and regulations as well as the control of both access to and output from the ocean.

State, Growth, and Marshall Funding

As indicated previously, the Danish state has played an important role in many aspects of both the growth and management of Danish fisheries. Since the latter half of the nineteenth century subsidies were given to the *captains of industry* in support of growth; and on the management level commission and laws were passed to shape and control the local activities. On an international level the Danish state took part in both the development of the EEZ, TACs and subsequently in the annual negotiations. Throughout the period described here, from the late nineteenth century to the present, the state has had shifting interests in the fisheries sector, from its potential to generate taxes to its role in economic growth and influence in the international race for fish. The first centrally administered Danish fisheries law, from 1888, was meant to ensure that operators paid a fee to the state, not to safeguard fish stocks (Vaarning 1984). This was the case with most legislation in this period, which was primarily concerned with defining the size and exemptions of the "sand tariff" paid by fishers (Vaarning 1984). The establishment of the Danish Royal Fishery Bank in 1932, as a credit institution for fishers, perhaps signals a shift to a growth perspective whereby the state mission changed from taxation to stimulation of the economy and focused on raising production in order to provide the population with proteins (Knudsen 2008). In 1947 came the first exclusive ministry for fisheries, which followed a liberal line of politics in order to create the best context for growth of the industry. In the 1950s the coastal inspection vessels even helped by using their fish finding equipment to localize schools of fish for the fishers (Søndergaard 2004).

During this period of "modernization" the fisheries sector became the target for intensive development work by the government. In the period from the mid-1950s to the mid-1960s, a total of seven laws created subsidization systems with the purpose of enhancing the development of fisheries. The first of these included money from the Marshall Plan. On top of this came the 1957 tax law on deductions that introduced tax exemptions on investments. A common thread through these legal changes was to increasingly promote the largest vessels and investments in volume. The subsidies were directed at technological change and at the need for the Danish fleet to be competitive in a global market. The important role of the state in actively stimulating growth in the sector is often left out of the narrative of open access tragedies, in which the lack of property rights and human nature are the only factors in the growing tragedy.

Besides stimulating economic growth, the state also had to promote its interests in international waters. This marks a shift whereby growth was increasingly funded to allow for participation in the international race for fish and the establishment of *historical fishing rights*. As mentioned above, similar intentions were already established by the end of the nineteenth century, with an escalation of competitive activity in international fishing grounds taking place in the second half of the twentieth century. An example of this is the Danish vessels "test fishing" in the North Atlantic for new and unexploited species in the 1970s. In 1983 national funding was supplemented with EU structural funding. At this point, the development of the fishery became much more complex as a double strategy with the aim of both modernizing the fleet and restricting its size was introduced by the EU. This problem was magnified by the closure of international waters caused by the introduction of EEZs and the subsequent return of distant-water vessels to their home territory (Holm 1996; Søndergaard 2004, 2008). In this twofold strategy, older vessels were taken out of operation through scrapping schemes while newer modern vessels were subsidized. This growth in the sector complicated the organizational unity of fishers, which had already been fragmented by differences in geography, gear, and political orientations.

Organizational Fragmentation

Almost as a precursor to later organizational fragmentations, the first attempts to develop a national fishermen's organization for Denmark in 1884 were subject to intense struggle. At a very late moment in proceedings, the organization's structure was changed to create an autocratic board (Hjorth Rasmussen 1984). As a result, an alternative organization was quickly initiated with greater member participation and more accountable local boards. The two merged in 1887, with the new overarching organization tasked with developing the Danish *fisheries in scale and profitability*. The main instrument for this development was to be the integration of science and fishing, the subsidization of enterprise and, on a practical level, the improvement of distribution and better handling of fish products (Hjorth Rasmussen 1984).

Having one organization representing the diverse and widespread Danish fishers was, from the beginning, a difficult task. There are plenty of areas where fishers could have conflicting interests: the promotion of gear types, fishing seasons, location of new harbors, relations with foreign fishers, favored export markets, shape of subsidies, to name but a few. The crucial role of the Danish seine technology in the expansion of Danish fisheries to the North Sea cannot be understated; but the technology also split the Danish fishers into two interest groups: active and passive gear users. The first fisheries law from 1888 reflected this fragmentation, only allowing Danish seine to be used inside the three-mile belt between the 1st of September and the 1st of April (Hjorth Rasmussen 1984). Growth in vessel size also created a division between the coastal fishery and the cutter fleet. The latter were too big to be hauled onto the beaches and were operated from the harbors. Later, the installment of motorized winches changed this aspect again and allowed larger cutters to be operated from the beach, though size continued to be a factor that divided the fleet.

In 1934 a large portion of the fishers from Western Jutland broke away from the national fishermen's organization and formed their own. This sectoral split occurred when Britain, in 1933, limited its fish imports from Denmark to 20,000 t annually. The British market was important, and many of the seagoing vessels from Western Jutland landed their catches directly to Hull or Grimsby. The problem was how to split this export "quota" throughout the year. The large vessel operators from Western Jutland wanted to split the limit on British exports into two six-month periods. With these longer periods, there would be less risk of being denied access to land fish in British harbors because a monthly limit had been reached. The smaller operators, however, wanted the limits to be monthly (Tarbensen 2012). They preferred the option of spreading out their export potential to have more equal access to British markets. As the latter were in the majority, it was this action that the national fishers' organization recommended to the government (Tarbensen 2012).

The internal differences in size and interests within the fleet led the overarching Danish fishers' organization to split into two separate institutions (Dansk Fiskeriforening and Danmarks Havfiskeriforening), a break which lasted until 1994. On the local level most fishers, both vessel owners and crew members, were organized into local branches of one of the two main organizations. These local organizations offered practical help for local fishermen and organized activities at the local level. The general principle was that the members paid a percentage fee on their income from fish landings. The local organization would then take care of many of the subsidiary fishing industries, for example, ice provision and distribution and packing, thereby bridging the gap between fishers and onshore distribution.

Small Craft and Trash Fish

Even though the Danish fisheries sector grew in size and complexity during the first half of the twentieth century, it was still primarily characterized by small-scale share-organized units. In 1948, a British journalist reported on the state of the Danish commercial fishing fleet for the trade paper Fishing News. The article noted that most Danish vessels were operated by an independent skipper and a small crew, and that Denmark only had ten vessels that were larger than 60 t. Among these larger vessels were the first steel trawlers, which were mainly employed in fishing activities around Iceland. The article was titled "Quality Comes First: Danish Fishing Not Based on Mass Hauls," pointing to a peculiar characteristic of the Danish commercial fishing fleet of the time.[11] Despite being equipped with smaller units than many other countries in the region, the Danish fleet delivered not only large quantities of fish but also fish of a high quality, a fact that the author attributes partly to the widespread use of Danish seine instead of trawl.[12]

In the second half of the twentieth century the Danish fisheries sector grew exponentially (Fig. 2.3). Catches increased tenfold from around 166,000 t in 1945 to almost 2 million t in 1980 (Holm 2002). A significant aspect of this rise in catches was the establishment of the industrial reduction fishery, known in Denmark as the "trash" fishery. Industrial reduction of fish began as an offshoot of fishing for human consumption, as bycatch was reduced to fishmeal and oil. With advancements in the fish meal processing technology, it was easier and cheaper to split the fish into oil, water, and dry material. Soon vessels were being designed purely for the purpose of catching fish for reduction. By the end of the 1940s, and with increasing rapidity in the 1950s, trash fisheries developed in Denmark. These were so important economically that the national fishermen's organizations campaigned to change the name of the fishery from "trash" to the less negative "industrial fish" (Tarbensen 2012). New species were utilized, and landings of these grew to become a substantial part of total landings and fishing pressure on the marine environment. In 1965, "trash fish" like sprat, Norway pout, and sand eel made up around 1% of the total landings in Denmark. By 1975, this number had grown to 60%. At the

---

[11] Fishing News, August 7, 1948, Page 3.

[12] "Typical too of Danish fishing is the use of the seine in preference to the trawl. Danish fishermen have specialized in this method, which is admirably suited to conditions in Danish waters. The seine does not damage the fish and often has a live catch, thus ensuring fish of good quality." Fishing News, August 7, 1948, Page 3.

Fig. 2.3 In the second half of the twentieth century, the Danish fisheries experienced both a substantial growth and decline leading to the introduction of an output quota. Many harbors that thrived in the 1960s and 1970s are dominated today by leisure activities, marinas, and tourist housing. (Photo: Jeppe Høst)

same time, the demersal fishery for consumption was also growing in landings and in fishing capacity, as were the pelagic fisheries for mackerel and herring. While the number of people directly employed in the fishing sector more or less stayed stable throughout this period—shifting from 15,000 to 20,000 between the beginning of the century and 1975—fish landings and the vessels under fishermen's feet grew substantially in size (Holm 2002). The so-called "second industrial revolution" of the fisheries took place based on new materials (most importantly nylon for nets), hydraulics, and electronic equipment (Søndergaard 2004) as well as the state financing described above.

The Introduction of Output Management

Record catches at the end of the 1960s, especially of herring, marked a paradigmatic shift in the management of Danish waters and in particular the problematic North Sea (Karlsdóttir 2005; Søndergaard 2008). Record herring catches of 1968 were followed by a 30% drop the following year. In a late 1970 edition of the weekly Danish fisheries magazine, a biologist made a prediction that set the scene for the remainder of the decade:

> We have to acknowledge that the free fishery of the ocean could soon be over with. There are simply too many well-equipped fishing vessels in the North Atlantic area. Right now there are international negotiations on the establishment of quota systems and there are a lot of indications that the fishers of the future will have to be licensed to fish. In return, they will be able to count on catching something. (Dansk Fiskeriforening 1970; author's translation)

Step by step, species by species, the quota system was introduced first by the North East Atlantic Fisheries Commission (NEAFC) and then, since 1977, by the European Economic Community (EEC), based on the scientific advice from the International Council for the Exploration of the Sea (ICES). Management of North Sea herring proved to be a trial run for what followed for other species and in other areas. First, time restrictions came on the fishing season in 1971. In 1974 quota were applied, and by 1975 a total ban on herring catches had been imposed. In the Baltic Sea, a commission was appointed in 1975 to distribute TACs between coastal states, a process that was complicated even further in this area by Cold War geopolitics. For both the Baltic and North Seas, the guiding principle was that distribution clearly had to consider previous fishing patterns, recognizing the primacy of historical catches. In this way, the 1970s could be considered as marking a departure from free and open access with the introduction of scientific management instruments. Indeed, the governability of the fisheries increased with the introduction of TACs and national quotas. It is important, however, to recognize that prior to this the sea fishery was not unregulated. What was essentially new in the 1970s was the regulation of output: the introduction of a national policy that aimed to manage the total biomass taken out of the sea.

Originally, the two Danish producer organizations had separately argued for two different solutions to the fisheries resource crisis. The organization mainly representing the largest operators argued for some sort of quota system, since a temporary closure would halt production from the very specialized units in the industrial fishery. The other producer organization, with a more mixed membership in terms of size and geography, preferred a temporary closure, giving the fish stocks time to recover. These units could fairly easily shift effort to other species. As we know, the 1970s led to the introduction of national quotas and TACs. Alongside the quota system, a new problem has come up for the management: how to distribute the limited quota. It would be erroneous to conflate the introduction of quotas with the invention of fisheries regulation. There had been plenty of regulation and control in earlier eras, but apart from minimum landing sizes this had mainly been on the input side—especially in regard to gear and fishing seasons. Ever since the end of the 1930s, for instance, mesh sizes and minimum catch sizes had been negotiated on the international level for the North Sea. After the Second World War this process really took off and was one of the reasons for establishing the Danish Ministry for Fisheries in 1947.

But property rights had also been used to manage fisheries. An example of this was the *eel yard right*, which gave farmers with property adjacent to the coast the exclusive right to set fish traps as far *as a pole could reach the bottom*. As such, the eel yard right was a property right and could be leased to the one doing the actual work. It was subject to disputes, and from the 1930s the Danish Fishermen's organization actively worked on its termination, with success in 1956 (Tarbensen 2012). One of the advocates of the eel yard right was Jens Warming, who in an article published in 1911 preceded the famous articles of Gordon and Scott with a similar argument for instating property rights and licenses in order to collect "sea rent" (Gordon 1954; Scott 1955; Warming 1981). Even earlier, the feudal system had imposed restrictions on the use of maritime resources. In the area around Hirtshals, for example, tenants had to deliver fish to the manor Adelsgård in exchange for the use of land and sea (Tophøj 1976). Such regulation was not uncommon around Denmark. Fisheries were regulated through existing social institutions, user rights at beaches, taxes, and citizen status. As in the surrounding countries, the right to fish on the open sea rested with the monarch, who could distribute it to warlords, peasants, or others as he deemed fit. In inland lakes, the rights over fish most often belonged to the manor (McCay and Acheson 1987).

That leads us to an important point. The free, equal, and open-access ocean of the twentieth century was something that the coastal population had fought for, with the expectation that it would later be protected by the rising democratic welfare state.[13] This open access—rather than being the result of a lack of management—was a domestic social institution guided by principles of equality. In similar lines, it can be argued that the open access structure presumed by Hardin, Gordon, and Scott to be primitive and universal was actually, in the "new world," the result of an independence gained from the old European world (Gordon 1954; Hardin 1968; McCay and Acheson 1987; Scott 1955). This historical regulation through limited access to maritime resources was in the eighteenth and nineteenth centuries slowly replaced by input regulations defined by the state. Gear types, mesh sizes, closed areas, and defined seasons were the guiding management principles with the main objective of protecting juvenile fish. Trawl was subject to many debates and banned from time to time, but often allowed again because of its widespread (illegal) use (Monrad 1997; Rasmussen 1968; Østergaard 1984). On the output side, minimum catch sizes were the main instrument to protect juvenile fish, all in order to maintain well-sized adult fish populations.

The primarily output-based management introduced in the 1970s did not replace but supplemented existing input and output regulations. For fishers, the introduction of quotas meant a new set of restrictions on their freedom. These new restrictions were decided at a political level and subject to negotiations both internationally and nationally. As they took place out of reach of most fishers, they significantly weakened the power of producer organizations and especially of individual fishers (Holm 2001; Søndergaard 2008). Output management was now based on scientific models strongly embedded in state bureaucracy (Hersoug 2005; Holm 1996) rather than, for example, co-management with the sector. New scientific advances, such as the Beverton-Holt model from 1957, improved the ability to estimate future fish

---

[13] "After all, it was with a particular social welfare function in mind that our founders determined that certain natural resources would remain the common property of all—not the private property of the few." (Bromley 1982, quoted from McCay and Acheson 1987, p. 195).

population sizes based on the knowledge of previous years (catches). The Beverton-Holt model was further improved by the virtual population analysis (VPAs) of the 1960s. These and other scientific improvements made it possible to mathematically offer estimations on the size and composition of fish populations and provided substantial background for calculating TACs (Rozwadowski 2002; Karlsdóttir 2005). The most important decisions in regard to quota sizes were now taken by scientists and governments, which placed the producers at the bottom of the governance hierarchy. It was at this time that Danish fishers went from being *captains of industry* to *clients of science*. Being subject to science-based output management meant that fishers more and more became "clients" in the system. As clients, fishers had to claim and apply for licenses, take care of a growing amount of paperwork and find their way through the rules of the system (Hersoug 2005; Søndergaard 2008).

The Quota Distribution Problem

On the management side, the introduction of the quota system in the 1970s also brought a new distribution problem. With a large fishing fleet and a new limit on the total output from the ocean, managers were faced with the problem of how they should distribute the right to fish between individual fishers and vessels. The use of quotas as a limit of the total output had initially been perceived as a temporary solution, only to be employed until stock situations had improved. But the quotas stayed and became permanent, and to solve the distribution issue the total quota were split into monthly limits. This system favored the largest units that could fish the largest quantities fastest—and in the worst kinds of weather. When the total quota had been "fished up" in one area, those boats with the capacity to work distant waters could move on to the next. This led to drastic changes in regional dynamics and seasonal mobility. These new behavioral consequences showed up only a few months into the new quota system and in particular divided trawlers and gillnetters into two interest groups. As a consequence, fishing units with a slower pace and an operation focused on quality and higher prices were slowly marginalized economically. Many of these quality-focused vessels had to consider if rigging the vessel to trawl, or even investing in a larger vessel, might not be the best option. One example of this was the wreck fishers of Hvide Sande, on the coast of Western Jutland. In the 1950s many of the vessels in Hvide Sande changed their gear from gillnets to trawl, but a large portion of these changed back again in the following decade (Moustgaard and Damgaard 1974). The reasons for this were high prices for premium quality fish, in particular cod and sole, and the adoption of new technologies that enabled fishers to locate and set their nets over shipwrecks. These fishers were active in adopting these technologies, originally developed for use in large-scale industrial vessels, to fit their smaller cutters. Sonar and echo sounders as well as maps and local knowledge were used in innovative ways to identify shipwrecks in the North Sea. On top of and around these wrecks, cod was abundant and of a great size, partly because the wrecks had to be avoided by bottom trawlers, and partly because the wrecks worked as an artificial "reef." In 1967, a local smith in Hvide Sande developed a hydraulic net hauler that, in combination with nylon nets, allowed the fishers to operate at much greater depth and all around the North Sea. Interestingly, a shallow sand bar outside the harbor placed an upper limit on the size of vessels landing there, and only a few were larger than 50 gross tons. These limitations on size forced the gillnetters to innovate and specialize. On land, new networks of roads and railroads served their specialized fishery, in combination with a local fish exporter with markets in European countries such as Spain, France, and Italy.

This example from Hvide Sande illustrates the complex interplay of technology, infrastructure, international markets, and local entrepreneurship in the development of a specialized fishery. However, it also illustrates a type of fishing that did not survive the introduction of output management. Today, fishers identify the introduction of national quotas as the end of specialized wreck fishing (Personal conversation). With a common quota for all fishers in the North Sea, gillnetters—and the wreck fishers in particular—lost out to larger trawl-rigged vessels that could fish up the limited quantities before smaller vessels and other gear types had the chance. This first phase of output management therefore required further refinement. In 1979, a new law passed by the Danish government granted the fisheries minister the right to administer and distribute the total quotas and to issue licenses—in other words, to limit the number of participants. The law also gave the minister the tools to create the ration system, which in time became increasingly advanced. But the minister's toolbox was, in principle, limited to restricting the quotas by time, catch areas, vessel categories, and licensing.

Two Decades of Rations

The system of output management then developed into a ration system, whereby annual quota was divided into monthly or fortnightly individual vessel rations. These differed according to a vessel's size and activity. To keep a high ration the vessel needed to show high activity. Inherent in this system was an incentive for fishers to go out to sea on the first day that the grounds were opened, regardless of the weather conditions, catch quality or price. If fishers caught their fortnightly allowance in a shorter time, they could sign up for another area or species (Frost et al. 2005). Alternatively, they would simply have to stay in harbor. The system distributed rights on equal terms to all registered fishers, but this allocation was increasingly complicated for managers. The dynamism of the system required constant recalculations to assess remaining quotas and resulted in problems with illegal landings (so-called black and grey landings)[14], which decreased trust in the system among fishers (Byskov 2007; Vedsmand et al. 1996; Vedsmand 1998). The absurdity of the ration system was clearest during the years in which TACs were set particularly low. When

---

[14] The illegal landings were primarily achieved by registering the catch onto another vessel, by registering it as another species or by changing the dates (to the next ration period).

these small rations were combined with low auction prices and rising oil costs, hardly anyone could survive from fishing. Large numbers of boats moored in the harbors were a visible manifestation of the overcapacity and distribution problem.

A series of protests in 1987 illustrates the problems in the sector. Fishers demanded compensation for all quota reductions since 1983, better scrapping schemes, a halt on import of fish products from certain countries, a debureaucratization of the regulations and a share of the income from state issued oil licenses (Tarbensen 2012). In 1993 came further quota reductions, which were followed by more fisher protests when quota rations in the Baltic Sea and Kattegat were limited to just 1 week and, in 1994, when trawling was suddenly banned in the Baltic Sea after mid-April. These disputes were calmed with compensation and new scrapping schemes. From a management point of view, the 1990s were characterized by the hope that state and EU funded reductions in the fleet would lead to a greater balance between fishing capacity and fish resources. However, from 1999 to 2002 TACs were reduced by up to 75% in some of the most vital catch areas in the North Sea. The combination of rations and scrapping was not able to solve the overcapacity problem, and the crisis of North Sea cod had political implications. The 2002 reform of the Common Fisheries Policy introduced a "days at sea" regulation, limiting the number of days a vessel could spend at sea. This was introduced for the North Sea and Skagerrak, but it had consequences for the whole country because the limitations led fishers to move away from these areas to fish in others. Over time, the way was paved to negotiate a radical break with the ration system and principles (Byskov 2010).

Diverging Solutions, Groups, and Interests

Debates as to how to distribute and manage limited outputs were ongoing even at the time of the introduction of the TACs in the 1970s and the establishment of the first national quotas. A key factor in these debates was the lack of unity both between and within the producer organizations. On many occasions this resulted in stasis or unilateral decisions made by the Ministry of Fisheries on behalf of the fragmented sector (Søndergaard 2008). Some sort of system for annual allocations had been on the agenda as early as 1981 but was removed because of segmented interests in the fleet. The largest operators were not interested in giving up their chance to catch more than others; the midsized vessels were afraid of losing the flexibility to switch between gears and species; and those in the near-shore sector who would have benefitted most were too weak to push their preferences (Søndergaard 2008). In the summer of 1989, discussions over distribution were once again on the agenda. Now the question was how to achieve sustainable development for the fisheries. The industrial fishery for pelagic species was now under strict control, and several of its cutters had switched to the consumption fishery. Increasingly, the public image of the industry was of a fleet too efficient for its own good (Pedersen 1990).

At this time there were three concrete options on the management agenda. The first was a fee system based on the idea that the least cost-effective operations would quit fishing as fees rose when quota were about to run out. Second, a system of licenses was proposed whereby annual amounts would be allocated to each operation. This would achieve stability in the sector, as each operator would know the annual amount of quota that he or she had to work with. Finally there was a similar proposal to this license system, but with transferable licenses and allocations. The organization representing the majority of large vessels was in favor of the latter proposal, claiming that it would have all the benefits of a license system with none of the disadvantages. Transferability allowed a higher degree of flexibility to lease quota in and out. The unions and fish exporters also promoted this option (Søndergaard 2008), most likely in an effort to stabilize the sector.

As described above, lack of sectoral unity had led step by step to an advanced system of rations, in which portions of the annual quota were distributed around the year in monthly, fortnightly, or weekly rations. The rations were not however distributed equally throughout the year, but changed in size according to the seasons. There were several good reasons for this: seasonal changes in catchability and spawning as well as fluctuating prices meant that in some months it was reasonable to have larger rations than others. This distribution was however also subject to political interests, especially the conflicting interests of place-bound fishers and larger operations with higher mobility. Fishers' numbers on the ground did not always match their influence in the organizations. One place-bound fisher explains in 2005:

> In this area, as the fishery is run now, it will collapse in a few years. It is under too much pressure, because of the politics in the organization. They work for high quotas in the first three months of the year; so all can take part in the fishery and then move on. The consequence is that the fish is taken at way too low a price. 90% of the fishers in this area are actually in favor of a system of annual amounts, but the fishermen's organization is against it. They work for the large vessels, which can go out in rough weather and go to other areas later. (Andersen and Andersen 2000, p. 65; author's translation)

The ration-based system was more than just an environmental management instrument. It was also a context for political conflicts and diverging interests within the fleet. The tools of this system provided compromises between the different interests but not lasting solutions. The economic situation worsened as investments in the sector grew and economic returns declined. Pressure rose for a new management regime, the form of which was constantly debated in both the fishers' organizations and the ministry (Søndergaard 2008).

The Arrival of Market-Based Fisheries Management

Individual non-transferable licenses had been on the agenda since the 1980s but did not have the support of fisher organizations, which argued that individual licenses would reduce the flexibility of the sector. Large operators opposed individual licenses because they would take away their chance to access more fish than other vessels (Søndergaard 2008). In the 1990s the Ministry of Fisheries shifted tactics and, instead of promoting universal management changes for the sector, initiated experiments in small parts of the fleet (Søndergaard 2008). In 2003, ITQs were introduced as an experiment in the herring fisheries. Based on a historical reference period, fishing rights were given as private property to vessel owners; and with these in their hands they could distribute the rights through buying and selling. In other words, market mechanisms were introduced, replacing public authority as the principal distributor of fishing opportunities. A few years later the experiments with herring fisheries were extended to the mackerel and industrial reduction fisheries, and they were finally introduced in the large and complex demersal consumption fishery in 2007. At the same time, the systems were made quasi permanent.[15] It is the introduction and functioning of the demersal management system, the VQSs, that is the focus of the remainder of this book.

In a matter of a few years, a large wave of market-based fisheries management washed over the Danish coastal areas. Fishing privileges based on equal access were transformed to transferable commodities based on the principles of individual private property. The allocated fishing rights could be used as collateral for investments in further fishing rights. This was an important aspect of the new system. As one fisher explains:

> We did not ask for it, the VQS system. It was the politicians. But then we could see, that was what made it possible for the Danish fishery to survive. Had we not gotten the VQS system, Danish fisheries would have died out slowly and surely. No one would have been able to offer anything. Had the ration system continued, then Danish fishery would have been dead. (Personal conversation, December 2012)

One of the biggest problems with the overcapacity of the Danish fisheries sector was the lack of a functioning market for vessels and fishing operations. A large group of vessel owners had entered the fishing sector in the 1960s and 1970s and was now trapped with their (over)investments because the fisheries were more or less closed to newcomers. At the same time, scrapping schemes were becoming more and more of a faux pas in the political climate, both at home and in Brussels. A positive aspect of market-based fisheries management for managers is that it can to some extent be seen as a privately financed scrapping scheme. With fishing rights as collateral and with the help of the financial system, suddenly some fishers had the opportunity to buy out the vessels of others in order to acquire their fishing rights.

Captains of Finance and the New Regulation

With the VQS system, vessel owners were suddenly freed from their status as clients of science and could reemerge as captains, but this time *captains of finance*. Vessel owners now had a new freedom to invest in quotas and shape their own fu-

---

[15] Technically the systems can be rolled back with 8 years notice. However, most people in and around the sector agree that this clause is unrealistic and impracticable at the current moment. Bank loans, for example, can be given on a 20-year term.

Fig. 2.4 The new regulation allowed operators to invest in fishing rights and to reorganize their operations and vessels to fit the new amount of quota. The photo shows a vessel being rebuilt and enlarged after having accumulated quota from other vessels. Photo: (Jeppe Høst)

ture. "The VQS system is the biggest revolution in Danish fisheries in more than 20 years," said the responsible minister at the introduction in 2006 (Ministry of Food and Agriculture 2006). What the revolution meant for the vessel owners was basically a new kind of freedom. Market mechanisms were introduced as a distributor of fishing opportunities; and those who invested experienced a new type of freedom, as they could be the active agents of their own enterprise (Fig. 2.4). As we will learn in the following chapters, some felt the system was morally wrong, and they quickly sold their operation or continued on leased "fish" in order to avoid buying quotas. Fish resources had suddenly been transformed from communal state owned goods into private property of vessel owners.

The VQS system was introduced as part of a political reform called the "new regulation" by the Danish Government. In contrast to the conventional understanding of regulation as "a diversion from what otherwise would occur, a blocking off, restriction or alteration of the alternatives open to the subject" (Mitnick 1980, p. 2, here quoted from Hersoug 2005), the VQS system enabled a whole set of new actions for vessel owners. The VQS system introduced completely new options and freedoms for vessel owners, who could now trade with quota shares. A consequence of this was the freedom to eliminate competition, and through market mechanisms vessel owners could buy up quota shares in neighboring or distant communities. This market possibility created new interactions and relationships between distant vessel owners and fishing communities. The race for fish and for a decent share of the ration was replaced by a fiercely competitive race for fishing rights (Højrup 2007).

The VQS system therefore did not only put an extra emphasis on the market, it created a whole new market for a new individual commodity. This marketization was originally met with resistance from the majority of fishers, beginning when it was first proposed in Denmark in the 1980s. These fishers feared the new basis for the fishing industry under marketization and the irreversible impacts this imposed on the fishing sector. Here the resistance is summarized by the organization Living Sea Denmark, a Danish NGO and one of the most critical antagonists of VQS:

> There are good explanations both for why a majority of Danish fishers do not want ITQs, and for why a minority would like to have it. You see, the majority of Danish fishers will disappear in such a regulation, for the simple reason that there is not enough fish and that the majority will not have the necessary capital required to fish under individual transferable quotas. It will be a fight for all against all, and that is already before one is on the sea. And such a regulation will most surely favor the indebted businesses which have no other choice than to buy the necessary amounts of fish—at any price. (Living Sea Denmark 2002)

The internal struggles and power shifts that finally led to the introduction of market-based fisheries management is another story (Højrup 2007). The new commodity and the new market changed the way of life of fishers and vessel owners. It had concrete implications for fishers: how to plan fishing activities, make a career in fishing, cope with the relationship between changing fish prices and quota prices, and deal with a newly minted property that was suddenly worth much more than the rest of the operation put together. These challenges, and others, are the subject of the following chapters.

A Break with Equal Access

The aim of this chapter has been to provide a historical background for understanding the Danish VQS system. I have shown that the VQS system builds upon a long development of regulations and is rooted in social institutions, technology, state and international politics as well as global competition. At the same time the VQS system represents a break with an epoch of regulations that were based on equal and free access and later the state management of a common national quota. Providing such a historical background is not as innocent an activity as it may seem at first glance. My motive is to show that the dominant narrative constructed around open access, capacity growth, and rational human behavior is just one way to represent the development process in fisheries. It is an act of framing that serves a purpose, often to promote market-based fisheries management systems. My aim here has been to open this field for closer investigation and examination, to prepare for an in-depth analysis of the contemporary conditions of Danish fishers, and to show the internal complexity of a market-based management system. The first step has been to outline the historical development and prehistory of Denmark's market-based regime. The history of commercial fishing in Denmark cannot be reduced to an open access tragedy. Without neglecting or denying the importance of technological development, it is important to recognize that a range of other aspects have played equally important roles. Technological developments on sea were paralleled and sometimes preconditioned by infrastructure on land, urbanization, and the growth of the welfare-state politics and the expansion of global markets. The changes in technology and in particular in vessel sizes can be seen as technological and economic industrialization, while the modes of vessel operation are still marked by hunting characteristics: a dynamic and dispersed resource. At the same time, there has always existed a vibrant fleet of relatively smaller share-organized units. In addition, the shifts in fisheries regulations have not been passive environmental control mechanisms. They have had an active role in shaping the Danish fisheries, moving fishers from catch area to catch area, shaping working rhythms and changing the type of work.

With a more nuanced understanding of the developments in the Danish fishing industry, we can see that other outcomes were possible. The present-day management regime is not a deterministic result of an open access resource and the unavoidable self-interest in human nature. What would the doubling of the catch after motorization have been worth without infrastructure and markets to distribute and consume that larger catch? Would the enormous growth in capacity in the 1960s and 1970s have been possible without subsidies and tax exemptions from the Danish state? What political agenda were these subsidies part of? What if reduction fisheries had been banned? The oceans were also not unregulated. The growth and management of the Danish fishing industry in the second half of the twentieth century were preceded by a long history of complex development and experience of management in the fisheries. With this management, the Danish state had ensured a general free and equal access, and from time to time when necessary regulated this through commissions and input regulation.

Markets—the Problem and the Solution?

Growing markets for fish have been both a driver for growth and a source of intense competition in Denmark as well as globally. This process has enabled fishing to be a full-time occupation that generated income in remote communities, while also increasingly demanding technological developments, geographical expansions, and constant investments. When resources began to decline and entry to the sector was halted, the market for vessels and gear froze, and overcapacity meant vessels were worth less and less. The market had not been able to distribute vessels to a new generation or align the capacity of these vessels to the available resource. In a market economy, the result of this would have been individual economic loses, but in the Danish fishery this was distorted by scrapping schemes and the transferability of some capacity rights (kilowatt and tonnage). Here one could argue that it is not a state responsibility to help out these individual and corporate economic tragedies. However, in the Danish consumption fishery a new market for fishing rights was chosen as the strategy to solve the problem of overcapacity and distribution of the limited total allowable catch. The main tool for this was private property rights that could be used by the *captains of finance* to accumulate fishing rights through the financial markets. If that can be argued to be a neoliberal turn, then it marks the end of a liberal period where the state fueled expansive growth while at the same time protecting the fish resource as common property in order to preserve equal rights to employment for the coastal population. With that in mind, we can return to the opening puzzle about a long past but short history: while fishing has a long past, it has a short history as a liberal occupation.

References

Andersen, Maja, and Knud Andersen. 2000. *Fiskere om fisk og fiskeri: essensen af 77 samtaler med danske kystfiskere om økologisk fiskeri: afsluttende rapport om en landsdækkende interviewundersøgelse blandt danske kystfiskere*. Grenå: Landsforeningen Levende Hav

Byskov, Søren. 2007. *Viljen til fiskeri: Hvide Sande-fiskeriet 1982–2007*. Hvide Sande: Hvide Sande Fiskeriforening

Byskov, Søren. 2010. *Fiskeriet der forsvandt: eksempler fra et dansk fiskerierhverv i opbrud 1990 -2008*. Esbjerg: Fiskeri- og Søfartsmuseet.

Bøggild, Hansaage. 1983. *Fiskerne på Bornholm: 1883–1983, Bornholms og Christiansøs Fiskeriforening*. Rønne: Bornholms Tidende

Cardwell, Emma. 2012. Invisible fishermen. The rise and fall of the British small boat fleet. In *European fisheries at a tipping point*, ed. T. Højrup and K. Schriewer. Murcia: edit.um.

Finley, Mary Carmel. 2007. The tragedy of enclosure fish, fisheries science, and U.S. foreign policy, 1920–1960. http://nsgl.gso.uri.edu/casg/casgy07001.pdf. Accessed 29 July 2014.

Finley, Carmel. 2011. *All the fish in the sea: Maximum sustainable yield and the failure of fisheries management*. Chicago: University of Chicago Press.

Frost, Hans, Jørgen Løkkegaard, and Jesper Andersen. 2005. Forvaltning af det danske konsumfiskeri. Copenhagen: Fødevareøkonomisk Institut.

Gordon, H. Scott. 1954. The economic theory of a common-property resource: The fishery. *The Journal of Political Economy* 62 (2):124–142.

Hardin, Garrett J. 1968. *The tragedy of the commons. Science*. Washington, D.C.: AAAS.

Hersoug, Bjørn. 2005. *Closing the commons: Norwegian fisheries from open access to private property*. Delft: Eburon.

Hjorth Rasmussen, Alan. 1968. *Dansk fiskeri gennem 100 år*. Fiskeri- og Søfartsmuseet

Hjorth Rasmussen, Alan. 1984. *Vejen til Nordsøen...: det søgående snurrevodsfiskeris gennembrud i Nordsøen og Skagerak 1884–1903*. Hirtshals: Nordsømuseet

Hjorth Rasmussen, Alan. 1993. *Østersøen gav -: træk af Bornholms og Christiansøs fiskerihistorie 1880–1993*. Rønne: Bornholms Historiske Samfund; Bornholms Museum

Hjorth Rasmussen, Alan. 2000. *De skabte et samfund–: glimt af udviklingen i Vorupør gennem 150 år*. Thisted: Museet for Thy og Vester Han Herred

Holm, Poul. 1994. *Fiskere og farvande: tværsnit af moderne dansk fiskeri*. Fiskeripuljen. Esbjerg: Fiskerimuseet.

Holm, Petter. 1996. Fisheries management and the domestication of nature. *Sociologia Ruralis*. 36 (2):177.

Holm, Petter. 2001. The invisible revolution: The construction of institutional change in the fisheries. Norwegian College of Fishery Science, University of Tromsø, Tromsø

Holm, Poul. 2002. Sjæk'len, årbog for Fiskeri- og Søfartsmuseet, Saltvandsakvariet i Esbjerg. *Sjæk'len, årbog for Fiskeri- og Søfartsmuseet, Saltvandsakvariet i Esbjerg*

Højrup, Thomas. 1984. Det fiskende menneske. In *Bygd*, ed. H. Meesenburg, 2–31.

Højrup, Thomas. 2002. *Dannelsens Dialektik*. Copenhagen: Museum Tusculanums Forlag.

Højrup, Thomas. 2007. Haw. Kampen for en levende kystkultur. Thorupstrand Kystfiskerlaug og Han Herred Havbåde. In *Sjæk'len*, ed. M. Hahn-Pedersen. Esbjerg: Fiskeri- og Søfartsmuseets Forlag.

Karlsdóttir, Hrefna M. 2005. *Fishing on common grounds: The consequences of unregulated fisheries of North Sea herring in the postwar period*. Göteborg: Ekonomisk-Historiska Institutionen vid Göteborgs Universitet.

Knudsen, Ståle. 2008. *Fishers and scientists in modern Turkey: The management of natural resources, knowledge, and identity on the eastern Black Sea coast*. Studies in environmental anthropology and ethnobiology, vol. v 8. New York: Berghahn.

Living Sea Denmark 2002. IK- stor ståhej for ingenting. www.levendehav.dk. Accessed 15 May 2011.

Löfgren, Orvar. 1977. *Fångstmän i industrisamhället: en halländsk kustbygds omvandling 1800–1970*. Lund: Liber Läromedel.

McCay, Bonnie J., and James M. Acheson. 1987. *The question of the commons: The culture and ecology of communal resources*. Tucson: University of Arizona Press.

Meesenburg, H., and T. Højrup. 1984. Det fiskende menneske. *Bygd* Årg. 15 (nr. 1):2–31.

Ministry of Food and Agriculture. 2006. *New Regulation—Press release*. Copenhagen: Ministry of Food and Agriculture.

Mitnick, Barry M. 1980. *The political economy of regulation: creating, designing, and removing regulatory forms*. New York: Columbia University Press.

Monrad, Kirsten. 1997. *Kunsten at overleve: de nordiske landes fiskeri set i historisk perspektiv*. København: Nordisk Ministerråd

Mortensøn, Ole. 2004. *Danske havne: en kulturhistorisk oversigt*. Esbjerg: Søfartspuljen; Fiskeri- og Søfartsmuseet

Moustgaard, Poul H., and Ellen Damgaard. 1974. *Garnfiskere: organisation og teknologi i et vestjysk konsumfiskeri*. Esbjerg: Fiskeri- og Søfartsmuseet, Saltvandsakvariet

Nandan, S. N. 1987. The Exclusive Economic Zone: A historical perspective. In *The law and the sea: Essays in memory of Jean Carroz*. http://www.fao.org/docrep/s5280T/s5280t0p.htm. Accessed 3 April 2012.

Pedersen, Hans, f. 1990. *Fisk for fremtiden: om havet som losseplads og arbejdsplads*. 1. Aufl. Aarhus: Modtryk.

Rasmussen, Holger. 1968. *Limfjordsfiskeriet før 1825: sædvane og centraldirigering*. Kbh: Nationalmuseet

Rasmussen, B. K. S., and Alan Hjorth Rasmussen. 1972. *Vestkysten*. Nyt Nordisk Forlag

Rozwadowski, Helen M. 2002. *The sea knows no boundaries: A century of marine science under ICES*. Copenhagen: International Council for the Exploration of the Sea in association with University of Washington Press

Scott, Anthony. 1955. The fishery: The objectives of sole ownership. *The Journal of Political Economy* 63 (2):116–124.

Stoklund, Bjarne. 2000. *Bondefiskere og strandsiddere: studier over de store sæsonfiskerier 1350–1600*. Kerteminde: Landbohistorisk Selskab.

Søndergaard, Morten Karnøe. 2004. *Teknologisk udvikling i dansk fiskeri 1945–2000. Fiskeri- og Søfartsmuseets studieserie; nr. 16*. Esbjerg: Fiskeri- og Søfartsmuseet.

Søndergaard, Morten Karnøe. 2008. *Dansk fiskeri c. 1945–2005: teknologi, udvikling og forestillinger om kontrol*. PhD. Esbjerg: Syddansk Universitet

Tarbensen, Kenn. 2012. *For alle Danmarks fiskere: Danmarks Fiskeriforening 125 år, 1887–2012*. Fredericia: Danmarks Fiskeriforening.

Tophøj, Knud. 1976. *Fra Peder Rimmens tid: fiskeriet ved Hirtshals omkring 1900*. Esbjerg: Fiskeri- og Søfartsmuseet.

Vaarning, Ole. 1984. *Fiskeriet fra Kerteminde i 18. og 19. århundrede*. Kerteminde: Kjerteminde Avis.

Vandsted, Torben. 2004. *Fra smult vand til modstrøm: Hirtshals Fiskeriforening 1904–2004*. Hirtshals: Hirtshals Fiskeriforening.

References

Veblen, Thorstein. 1964. *Absentee ownership and business enterprise in recent times; the case of America*. New York: A.M. Kelley, bookseller.
Vedsmand, Tomas, and Bornholms Forskningscenter. 1998. *Fiskeriets regulering og erhvervsudvikling—i et institutionelt perspektiv*. Nexø: Bornholms Forskningscenter.
Vedsmand, Tomas, Peter Friis, and Nielsen Jesper Raakjær. 1996. *The Danish fishing industry,—structure, policy formulation and control of Danish fisheries*. Roskilde: Roskilde Universitetscenter, Institut for Geografi og Internationale Udviklingsstudier.
Warming, Jens. 1981. *On rent of fishing grounds: a translation of Jens Warming's 1911 article*. Staff paper series; 81–13. Kingston, R. I.: University of Rhode Island. Department of Resource Economics.
Wilcox, Martin. 2006. Concentration of disintegration? vessel ownership, fish wholesaling and processing in the British Trawl Fishery. In *North Atlantic fisheries: Supply, marketing and consumption*, eds. David J. Starkey and James E. Candow, 50–71. Hull: North Atlantic Fisheries Association.
Zeller, Dirk, and D. Pauly. 2007. *Reconstruction of marine fisheries catches for key countries and regions (1950–2005)*. Vancouver, B.C.: Fisheries Centre, University of British Columbia.
Østergaard, Jens. 1984. *Vodbinderi og fiskeri i Limfjorden: vodbinder- og fiskerihistoriske skildringer 1867–1982*. Nibe: Valsted.

Chapter 3
Society and Market

Abstract This chapter begins by discussing what a market is and examining the relations between society, state, and market. This discussion serves to broaden fisheries policy design as a social and cultural object of inquiry, seeing regulations as related to social groups and cultural forms and their agency. The chapter continues by examining the policy design of the Danish market-based fisheries management system. The chapter asks what is at stake when a market is introduced and further evaluates the concrete design of safeguards and anti-concentration rules. Rather than a best-case example, the chapter shows that the Danish Vessel Quota Share system is full of flaws and contradictions in its basic design.

Keywords Quota concentration · Trawling · Market economy · Social safeguards · Quota trade

Market Economy

The fundamental new feature in market-based fisheries management is the establishment of a market for trading fishing rights. In the previous chapter, we saw how this market subsequently became the principal distributor of fishing opportunities and activities. In this chapter, I will examine the vessel quota share (VQS) system as a market-based management system, with particular emphasis on the social dimension of the policy. How is the system designed to cope with a mixture of policy objectives? How can a system balance the needs of profit-seeking companies with societal and environmental objectives? Through this analysis, I will describe and explain the policy in detail, tracking the developments in quota transfers between 2007 and 2011. Before I begin this task however, I will ask a fundamental question: What is a market and what is a market economy? According to Karl Polanyi,

> Market economy implies a self-regulating system of markets; in slightly more technical terms, it is an economy directed by market prices and nothing but market prices. (Polanyi 1957, p 45)

Division of Labor and the Economic Man

Today, different markets are such an integral part of everyday life that most people hardly question them. On the other hand, the labor market is currently debated on a daily basis by politicians, the press, and the public in Denmark; global financial markets have recently caused worldwide recession; and inside the European Union (EU), we protect ourselves from market competition from outside economies. So perhaps, although markets are accepted as part of the everyday world, we have not yet come to fully understand the processes and implications of the market economy, where resources and opportunities are distributed through market mechanisms. With this in mind, I will use the next section of this chapter to question the definitions, and implications, of a market economy. A key aspect of this analysis will be undertaken through an examination of the arguments made by Karl Polanyi in his book *The Great Transformation* (first published in 1944). In *The Great Transformation,* Polanyi is concerned with the role of trade and markets in society and especially the implications of the growing market economy for labor, land, and money.

Polanyi undertakes a critical examination of economic history and targets some of the assumptions made by early classical economists, assumptions that have broadly influenced economic thinking since then (Graeber 2001). These can all be more or less traced back to one central assumption, namely Adam Smith's basic assertion that the "propensity to truck, barter and exchange one thing for another [...] is common to all men, and to be found in no other race of animals (Smith and Cannan 2003, p. 42). Much can be understood from this notion. Not only are humans involved in trade, this action is based on a rational capacity to exchange and barter. From this starting point, the individual is established as the central analytical figure, and the division of labor deduced from trade. According to classical economics, the division of labor (and with that the complexity of society) is rooted in the universal tendency for humans to rationally exchange goods. The person who makes "bows and arrows, for example, with more readiness and dexterity than any other" (Smith and Cannan 2003, p. 43) will focus on producing bows and arrows and exchange them for food instead of participating in the hunt himself. This behavior leads to division of labor. In other words, rational individuals will specialize in one trait and, through exchange, supply themselves with the necessary goods, increasing total production in society. As such, the division of labor is a function of exchange; and thus, the size and shape of the market has considerable implications for the rational division of labor:

> As it is the power of exchanging that gives occasion to the division of labour, so the extent of this division must always be limited by the extent of that power, or, in other words, by the extent of the market. (Smith and Cannan 2003, p. 45)

Therefore, in classical economics, properly functioning markets are important as they enable rational behavior. While the linkage between the maximization of trade by the individual and the division of labor became one of the basic ideas in classical economics, it also served as the foundation of the idea of the economic man—the rational, utility-maximizing individual (Rittenberg et al. 2008).

Thus, restrictions put on markets and trade in Western European societies were the main target for classical economists arguing for "liberalization" of the economy—what today would be termed "deregulation." Interestingly, Smith himself did not advocate a pure market without regulation, instead positing that checks and balances were vital to protect human welfare (Aguilera-Klink 1994). However, market-based fisheries management systems are constructed around the assumptions about the economic man described above. In fact, problems of fisheries management are said to derive from the natural and rational behavior of individuals and the lack of property rights—a problem framed as the tragedy of the commons (Hardin 1968). The following example, from an international review of individual transferable quotas (ITQ) systems, begins by defining human nature:

> When many fishermen have access to the same fish stock, each has every reason to grasp as large a share of the potential yield as possible lest the other fishermen reap all the benefits the resource can offer. (Arnason 2002, p. 1)

In order to avoid this race for fish, the consequent economic solution is to establish property rights, which will prevent overharvesting. In addition, a market for these rights will ensure efficiency. The best (or most efficient) fisher will have the highest profits and be able to offer the best price, so the market ensures that fishing opportunities are transferred to the most efficient users.

> In theory, a piece of property will be most valuable to the most efficient and farsighted owner. Thus, the sale of property to the highest bidder should place property in the hands of the most efficient and farsighted user. (Townsend and Wilson 1987, p. 312)

There are numerous other examples from the economic literature on market-based fisheries management that illustrate the reliance on the fundamental assumptions of classical economics and the tragedy of the commons (see, for example, Gordon 1954; Grafton 1999, 1995, 1996).

What Polanyi does, through the analysis of ethnographic and historical material, is to show that the division of labor exists in several societies and communities—even those without markets, and especially those without the economic motives of personal gain. These communities are instead organized through three other distributive principles: reciprocity, redistribution, and householding. In this way, Polanyi tackles the link between division of labor and trade, and he demonstrates that other principles can form the basis for complex and advanced economies and social formations. Through these three principles, goods and wealth can be distributed and shared without a market but with a division of labor.

Reciprocity, Redistribution, and Householding

For Polanyi, reciprocity, redistribution, and householding were three key principles that could be used in different combinations and extents to explain the division of labor in societies where markets played a minor or nonpermanent role. Since these three principles are important in Polanyi's account, I will briefly review Polanyi's arguments here. Reciprocity is a sort of gift economy, which, as Polanyi shows, can be rather advanced in terms of geographical reach and social complexity. *Reciprocity* works through the principle of symmetry, where "today's giving will be recompensed by tomorrow's taking" (Polanyi 1957, p. 51). However, this symmetry does not need to be a direct one-to-one relation. Instead, what one gives to his wife's family is recompensed by what he receives from the husband of his sister and so forth. Through such dualities between sexes, families, generations, villages, and islands, reciprocity can cover large geographical areas and social complexities—including the division of labor—without the principle of truck, barter, and exchange. The second principle, *redistribution,* is the circulation of produce to a central chief who keeps it in storage and redistributes it through communal activities and trading with other communities (Polanyi 1957, pp. 47–49). Redistribution works through the institutional pattern of centricity, which "provides a track for the collection, storage, and redistribution of goods and services" to a central institution or person (Polanyi 1957, p. 49). As Polanyi shows, redistribution is not confined to small economies but can provide the economic framework for much larger societies:

> Economically, it is an essential part of the existing system of division of labor, of foreign trading, of taxation for public purposes, of defense provisions. But these functions of an economic system proper are completely absorbed by the intensely vivid experiences which offer superabundant noneconomic motivation for every act performed in the frame of the social system as a whole. (Polanyi 1957, p. 48).

The redistribution principle described above is found variously in the ethnographic literature from small island societies of the Pacific Ocean to the larger feudal states of Europe. This is also the case with the third principle, *householding,* which is production for one's own use.

Householding is based on the institution of autarchy and self-reliance, as it is the type of economy within which the nucleus unit (i.e., family, the settlement, or the manor) produces most of what it consumes—and consumes most of what it produces. The key point is that this can occur with substantial internal division of labor (Polanyi 1957, pp. 47–49).

As mentioned above, Polanyi uses the three principles to contrast some of the basic assumptions in classical economy: the relation between trade and division of labor; the idea of humans as, in essence, economically maximizing individuals; and the universality of the market economy. Two points are worth highlighting in this context. First, Polanyi shows that the utility maximizing economic man is not a universal entity, but is instead related to the market economy and the market pattern (to the motive of exchange for profit). Thus markets and the rational behavior universalized by economists are linked to one another. The individual actions Polanyi finds and describes are in stark contrast to actions of the economic man but directed towards, for example, social status and assets:

> He does not act so as to safeguard his individual interest in the possession of material goods; he acts so as to safeguard his social standing, his social claims, his social assets. He values material goods only in so far as they serve this end. (Polanyi 1957, p. 45)

Instead of economic gain, Polanyi describes human passions directed towards noneconomic ends and production that is integrated into social life (Polanyi 1957, p. 45). This integration of economy and production with social institutions and social life, for which Polanyi is often referenced, is the second point I wish to highlight. In the societies Polanyi describes with the principles of reciprocity, redistribution, and householding, he explains the economy as *embedded* in social institutions. Polanyi's idea of an *embedded economy* is a conceptual contrast to economists' ideas of market economies, where separate economic institutions have been created and consequently "disembedded" from social life. Even though Polanyi is often credited for the idea of economy embedded in social institutions, it is not a coherent theory or conceptualization he lays out in *The Great Transformation*. Instead, Polanyi argues that views of the market economy as an integrated human function and exchange as a universal human tendency were merely the projections of recent developments during the time of the first classical economists (Polanyi 1957). The market as a dominant organizing principle is therefore not universal; rather, markets exist in many societies (in different times and places) but without inherently playing an organizing role in society. What was new in the late eighteenth and nineteenth centuries in Western Europe was the growing role of the market as an organizing principle; the market as a separate economic institution with its own laws, its own regulations, and the distribution of labor, money, and land through the principle of price. In short, what was new was the market economy.

The Market Economy and Social Safeguards

The markets we know today as the labor market, land market, financial market, and the commodity market were in principle created and regulated through state intervention (Polanyi 1957, p. 63). Rather than being a natural quality of human action, markets were established and shaped by politics. It is this state intervention and subsequent state effort to uphold the markets that Polanyi termed *the great transformation:* a transformation that places society as an adjunct to the market economy. Markets were created as separate institutions to distribute not only commodities but also land, money, and labor through economic transactions and individual contracts.

> For once the economic system is organized in separate institutions, based on specific motives and conferring a special status, society must be shaped in such a manner as to allow that system to function according to its own laws. (Polanyi 1957, p. 57)

In order for the markets to work, it was necessary to reshape society. Historically, guilds and traditional land tenure systems had to be abolished. This transformation, Polanyi argues, was opposed by a society that tried to safeguard its social organization against market behavior. These developments in the nineteenth century were, in his words, the history of a *double movement* where "the extension of the market organization in respect to genuine commodities was accompanied by its restriction in respect to fictitious ones" (Polanyi 1957, p. 76). On one hand, markets for goods were spread all over the world; while on the other, state policies and civil society organizations were created to check the effects of the markets on land, labor, and money (what Polanyi terms fictitious commodities).

Until the Industrial Revolution in Western Europe, markets were contained as "accessories of economic life," and the economy was largely still embedded in social institutions and relations (Polanyi 1957, p. 68). But with the formation of new markets for fictitious commodities came also a new set of regulations and institutions to safeguard social life. Labor unions, philanthropist organizations, welfare institutions, health care, and social security complied with a substantial body of law and regulations as a balance to the market economy. In Polanyi's account, the market economy came about as part of industrialization and the need to expand industrial production. The invention of "elaborate and therefore specific machinery and plant" (Polanyi 1957, pp. 74–75) changed the relations of production and were the historical push for the creation of the market economy, the formation of markets for labor, land, and money. The money commodity was necessary to raise capital for the large investments in machinery, the labor commodity was necessary to plan and control production, and the commoditization of land was a precondition for investments to be sensible in the long-term. It is in this way that we know these commodities by their commodity price names: wage, interest, and rent.

The Market Economy and Fishing Rights

Recalling the enormous technological development and growth in the fisheries sector during the twentieth century, it is interesting that Polanyi connects the evolution of the *market economy* with the industrial growth and mechanization of production. As I have shown in the previous chapter, the average size of fishing vessels and the amount of technology involved in fishing increased in the twentieth century. The ratio between the number of men on board and the amount of technology applied to catching fish became more similar to a factory than a simple production unit, a guild, or a family. The question is if the marketization of fishing rights should be understood broadly as part of this industrialization process of the fisheries sector. On land, industrialization and state intervention brought labor, money, and land into markets, a process that changed the *qualities* of these *commodities*. As Polanyi describes, prior to industrialization labor was a necessary condition of life, nature the basis of life, and money simply a token of purchasing power. This is why Polanyi terms these new commodities *fictitious commodities,* which—unlike genuine commodities—are not produced for exchange on a market. A consequence of this is that they are easier to monopolize, much like the best plots of land, access to credit or labor organized in unions. So too, fishing rights can be termed fictitious commodities, as access to fishing is not a genuine produced commodity. No one will produce new fishing rights for sale, and their ownership means an exclusive access to the benefits from the resource (Ribot 2003).

As was the case with labor, land, and money, the market for fishing rights was formed by state intervention in Denmark through a series of political decisions and the so-called New Regulation in 2007. As is well known, the formation of markets, fictitious commodities, and, more broadly, the capitalist system created new social relations, most importantly those between capitalists and the labor-selling classes. But this process also had to break down and transform existing social relations. Feudal land systems had to be dismantled and guilds transformed in order for labor and land to be available through markets. Economic institutions had to be established to govern and protect the use of money. In Polanyi's account, society and social institutions became adjuncts to the market economy, a situation Polanyi described as the economy being *disembedded* from social relations. No longer were economic actions rooted in social institutions. Such a situation where society is an adjunct to the market economy is, according to Polanyi, unstable and problematic, leading to the degradation of both nature and society. In this way *The Great Transformation* is not only a critique of assumptions made in classical economy but also of the economic model and development at the time of writing. Despite this, many of Polanyi's general and principal points still have analytical use and validity today.

The Market Economy and Social Science

For Polanyi, the introduction of the market economy and the commoditization of land, labor, and money is also the *raison d'etre* for political economy as a scientific field (Polanyi 1957). Understood in this way, political economy is concerned with a range of ethical, philosophical, economic, and practical issues around the distribution of wealth and resources, as well as with the relations between people in a market economy. At the same time, the disembedded economy functioning in a separate market institution also created an autonomous field for economics as a science, namely the modeling, forecasting, and political recommendation for intervention in markets. Perhaps this divide between society and economy also explains the differences between social sciences like ethnology and economics (Gudeman 2008) and can help explain the gap between the disciplines in the literature on fisheries management in general and market-based fisheries management in particular. Seen from this perspective, the two disciplines have completely different objects of research.

On one side, ethnology and anthropology, as well as other similar disciplines, are focused on communities and social systems with a multiplicity of values and social relations; on the other side, economics is concerned with the behavior of individuals and market patterns (Graeber 2001). In the latter perspective, relations between people are reduced to relations through a market. What is seen as negative concentration of quotas and decline of communities by one side is seen as rational specialization by the other; or perhaps as a reason to intervene to avoid the self-destruction of the market. In this respect, it can be argued that the introduction of markets for fishing rights helps create a regime for behavior that can be easier understood and analyzed using the conceptual apparatus of economy as a science, whereas it dismantles, or disembeds, the phenomena studied in other social sciences and humanities. However, this dichotomy is rather simplified and subject to the same critique as Polanyi's model of embedded and disembedded economies. The economy is socially organized, and economic choices are part of social life. The people involved in commercial fisheries are both individuals acting alone and members of larger social systems, networks, and complex value environments. In addition, economy as a discipline has its own divides and discussions around its own basic assumptions, which are much more complex than the simplified account given above. As I have pointed out, however, there is a strong link in the economic literature on market-based fisheries management and the basic assumptions described above. Polanyi's work should, of course, be used with care and critically examined before being reproduced. Below I will shortly discuss the critique of Polanyi's work.

Embedded, Disembedded, or Both?

In the previous section, I revisited the theories of Polanyi in order to understand the market institution from a historical and anthropological perspective. Polanyi's work gave rise to extensive debates in anthropology around the further elaboration of his ideas (for example: Beckett 2009; Gudeman 2008; Hann and Hart 2009). While many of Polanyi's general points still stand today, some of his arguments have been refined or dismissed.

One debate is concerned with the origin of the market economy. If *The Great Transformation* is a critique of classical economics and its economic history, Polanyi needs to offer an alternative explanation for the rise of the market economy more compelling than the deterministic argument of market economies as the teleological fate of *economic man*. Polanyi proposes an account centered on the active role of the state and the needs of the factory owner to organize and plan production. This interpretation has, of course, been subject to debate. Stephen Gudeman suggests that impersonal trade was the driving force behind the rise of markets (Gudeman 2001, 2008), while others use the notion of *douce commerce* to suggest that the peaceful character of trade was key to its growing importance since the late seventeenth century, as it contributed to stabilizing the political situation in Europe (Hirschman 1977). For this inquiry into the market-based fisheries system in Denmark, it is not necessary to fully engage in this debate. However, the critique of Polanyi's account often misses one important point, namely the active state role in supporting the markets. Like the markets for labor and land that Polanyi described (primarily in the English context), the market for fishing rights in Denmark was created by state intervention. The market for quota shares did not grow out of a universal human capacity to trade.

The strong dichotomy between embedded and disembedded economies has been much criticized (Beckett 2009; Gudeman 2008; Hann and Hart 2009). The academic consensus seems to be that even in a market economy, social relations and networks still play a large economic role, and economies are always both embedded (mutual, reciprocal, and based on networks) and disembedded (based on impersonal exchange) (Beckett 2009; Graeber 2001; Granovetter 1985; Gudeman 2001, 2008; Hann and Hart 2009). Perhaps Polanyi's strong dichotomy should be seen as part of his project to criticize classical economics, with pessimistic predictions for the future of the market economy (Hann and Hart 2009). It is possible, like Gudeman (2008), to see Polanyi's dichotomy rather as a dialectic relation between embedded and disembedded aspects of the economy, mutual relations (in groups and broader in society), and impersonal trade in a market. Instead of a dichotomy of two separate situations, seeing it as a dialectic relation stresses the connection and solution of two or more divergent objectives into one practice. In this dialectical relation a person is simultaneously part of one or more communities (with loyalties and mutual values) as well as an individual who can act rationally and seek maximum utility in the market. In this way, changes in the relation between the two, as well as in the actual specification of this relation in real-world terms, can be studied as a dynamic force with the individual subject navigating and situated between more or less embedded positions.

The quota companies established in the first few months of the VQS system are good examples of this process. In these, groups of individuals formed organizations that were opposed to the free market. Quota companies were group investments in quota shares, based on a certain loyalty or mutuality with their own and neighboring communities. At least three different principles were applied when the fishers formed these joint quota companies. One of the companies, in Thyborøn Western Jutland, was established as a joint stock company with 31 shareholders who would lease fishing rights from the company. The board determines the leasing price, and surplus quotas are leased on the open market. Inside the quota company, the market mechanisms have been blocked and are now based on ownership of shares. A second quota company established in central Denmark divided a common investment into equal shares. The fishers had to finance some of the investments themselves, and some hold more than one share. The prices of the shares can fluctuate, and they can be sold to others who wish to benefit from the investments. Each share gives access to a certain amount of quota as well as to the financial obligations to pay off the joint loan. After the initial distribution early in each year, there is another round of internal swapping of quota before surplus is leased to the open market. Here some degree of reciprocity is introduced, as the quota holders swap between them, sometimes even through the central institution of the quota pool manager. Finally, a third organization was formed in Thorupstrand Northern Jutland, here called a quota guild. Each person contributed 100,000 DKR as collateral to take a communal loan. New members can join by making a similar deposit. In this organization, the quota is distributed in equal shares over several rounds. The fishers have to lease the quota from the organization; when they decide they have enough, they refrain from accepting further quota. The remaining quota is then distributed to any fishers who require more. In contrast to the two other quota companies, fishers without vessels can also be part of this quota company. In addition, the guild uses the open leasing market to swap and lease excess quota. Without being straight-forward versions of Polanyi's principles—*householding, reciprocity,* or *redistribution*—it is, however, easy to see that the joint quota companies break with the market principle, each in a different way. These three examples are different technical solutions to the discrepancies between local loyalties and free market agency. Underpinning them are of course more complex social and cultural situations, which in turn shaped the distinct solution and model chosen. In Gudeman's words, this situation represents a dialectic relation between two "realms" (Gudeman 2008). The mutual realm could be described as a group or closer community, where the principles of reciprocity and redistribution are present together with other complex value systems, whereas the market realm is dominated by the market pattern—the individual maximizing utility from exchange. It is interesting here that the quota company itself becomes a player in the market, with the purpose of maximizing benefits for the group.

The complex dialectics between individual, group, and market were revealed when quota prices dropped after the credit crunch in 2008, and the quota obtained in common through loans moved into negative equity. The cases described here show how the market has become embedded in local contexts and also how new group structures are enabled by market-based fisheries management. The simple dialectic relation between the individuals and the market pointed towards this phenomenon, but only partially explain the formation of quota companies. For example, it takes for granted already existing group dynamics and the formation of a legal system, which allowed these new structures to be established. In Polanyi's words, the above examples could be seen as communities or groups safeguarding themselves from market mechanisms.

Safeguarding and Market-Based Fisheries

In Polanyi's account, the market pattern is essentially destructive for nature and society, and therefore he finds societies safeguarding themselves from the expansion of the market pattern. This is what Polanyi calls the double movement, with new and expanding markets on one side followed by legislation, organizations, and initiatives to moderate the negative impacts of the market on the other. This point actually undermines the strong dichotomy Polanyi originally laid out and promotes the idea of examining how markets and society are embedded in each other under new conditions. In his conception of society, Polanyi believed that "society was a natural form designed to provide material sustenance for its members" (Hann and Hart 2009, p. 4). Using this definition it is possible to establish the dichotomy between society and markets, but it is difficult to review its complexity and composite character. Society becomes what is inside this natural form from case to case, and it is not conceptualized in any systemic way. This conception of society does not provide insights into the structures society is made up of and how they relate to each other. This was not Polanyi's main project, and instead we learn about these through his historical accounts. For example, the safeguarding of labor, land, and money was a political debate between employer and employee, played out through labor unions and political parties as well as state assignment.

Polanyi's notion of society was used to cover wide historical and geographical spans. Had he only focused on the rise of capitalism, he could have used another notion of society, one which is strongly linked to the markets. *Civil society* is a historical and conceptual construct that parallels the market economy and the economic transactions outside state and family. It is this Hegelian understanding that is carried on in Marx's understanding of society as an economic base—including the struggles between the different classes in the capitalist economy. In society, interest groups, ranks, and classes organize and debate their conflicting interests. More than a natural form, society is made up of a plethora of interests and wills, with the state as a central factor. This safeguarding against the market is the outcome of opposite wills and interest groups and their political and ideological conflicts. Therefore safeguards can be part of a "built-in" design when the market is established and continuously regulated by the state. In other words, the double movement—and therefore safeguarding—is part of the formation of the market. This means that the market design already is a social product. This enables the social scientist to examine the social product in the policy design. Thus, in relation to the introduction of market-based fisheries management, the actual design of the system will in most cases encompass features to balance market mechanisms and societal objectives as well as environmental aspects. This design would represent the policy compromise between different wills and interests. Indeed, such features designed to protect certain segments of the fleet (i.e., small-scale fishers) are often called safeguards. In this understanding of society, it is a sphere where differing interest groups come to term with the powerful effects of the market on communities, environment, families, individuals, and so on.

The Economics of Market-Based Fisheries Management

The brief introduction to Polanyi's work and its subsequent use as presented above serves to contextualize an economic market as the object for ethnological investigation. Polanyi's arguments bring up several relevant points. Markets were shown to be historically specific rather than universal, and based on more than transactions of individuals. Instead, markets were also state created, maintained and supported. When in the following sections and chapters I turn to the description and analysis of the VQS system, I will use a number of analytical waypoints. One analytical waypoint is to understand and analyze how individuals navigate between the opportunities in the market and the loyalties and mutual values in social groups and communities. I discussed the potential of this analysis in regard to the quota companies and guilds earlier in this chapter. Another analytical waypoint is to examine how fisheries policies are formed and shaped by a multitude of interests, while governed and regulated by a state. By analyzing the presence of these different groups and interests in policy, I can move towards an understanding of market-based fisheries management as a social product and an outcome of complex societal relations.

Multiple Objectives

In recent years, the debate around market-based fisheries management has developed into a discussion about how to design market-based systems to avoid negative social and environmental impacts. Multiple objectives have to be balanced into the policy by design (Høst 2012). This is in line with the points made above about the social product in management and market policies:

> The goals for ITQ programs go beyond economic efficiency and encompass vibrant coastal communities, preservation of fishing communities and their culture, and healthy ocean ecosystems. Using a single tool developed to eliminate the economic waste associated with derby-style fisheries to accomplish all these goals simultaneously only increases the difficulties in the design stage. Trade-offs are inevitable. (Sanchirico and Kroetz 2010, p. 42)

As the number of active market-based fisheries programs increases, there is a growing amount of literature on the different designs of market-based systems. This literature, which is partly academic but also increasingly produced by civil society organizations, fuels a discussion of how to avoid negative impacts of the market through innovative design. According to one report, published by the corporate-financed environmental NGO Environmental Defense Fund in 2010, the Danish programs are examples of systems with innovative design:

> The Danish Pelagic and Demersal Individual Transferable Quota Programs (ITQ Programs) include a number of thoughtful design decisions in order to meet the programs' goals, including promoting economic growth in the fisheries sector by balancing the capacity of the fishing fleet with the available resource, and addressing social concerns. Important features of the catch share program include quota set-asides for small vessels and new entrants; Fishpools, which promote cooperation and coordination among participants; and programs to reduce discards. Denmark's catch share programs demonstrate how innovative design features can be used to promote social goals within a system introduced for economic and biological reasons. (Bonzon et al. 2010, p. 137)

According to the Environmental Defense Fund, the Danish programs address social concerns by setting aside quota for small vessels and new entrants, thus implicitly balancing out some of the negative societal impacts of market-based programs (and addressing values of intergenerational equity and diversity). Likewise, in the current 2012 reform of the Common Fisheries Policy of the EU, the European Commission on Fisheries uses the "Danish model" as a positive example, arguing in a briefing paper that it models a member state "where a TFC [Transferable Fishing Concessions] system is used shows that risks can be avoided through design." (European Commission on Fisheries 2012, p. 2)

The State and Free Markets

As already noted, Polanyi's economic history directs attention to the fact that in many examples, the state played a significant role in the creation and upholding of economic markets. In practice, the free and self-regulating market relies on the presence of a strong, effective state and a legal system to secure, guard, and regulate property rights. Likewise, the Danish VQS system was introduced as part of a total allowable catch (TAC) regime that calculates the total allowable output of the oceans. Here issues of overfishing are related to control and compliance of individual and total landings. The TAC regime is dependent on international agreements through the International Council for Exploration of the Sea, in addition to national control. A strict TAC (or a similar regime) is a precondition for a market-based system such as the Danish VQS system or ITQ systems in general.

But control and regulation is not the only way the state played an important role in the creation of the new market for fishing rights. The introduction of market mechanisms to distribute fishing opportunities between operators was created through state interventions. In practice, it was the dual introduction of both a new commodity that can be transferred through market mechanisms and a legal framework to facilitate and regulate this trade. The commodity introduced was a *harvest or fishing right,* and in this specific case these were named VQS, which were given as private property[1] to individual operators following the principle of historic catch. The dual introduction of a legal market (transferability) and quota shares (commodities) initiated a process of transferring, leasing, and swapping fishing rights between operators (vessel owners). As we will see in the following chapters, this process quickly became commonplace, taking place at harbors, through shipbrokers, on the telephone or via the Internet in one of the "fish pools."

Transfers were carried out between companies and individuals, neighbors in the same community, or people completely anonymous to each other through specialist brokers and accountants. With the introduction of a market the future distribution of marine resources is now mainly organized by the principle of price, governed through contracts between companies and individuals. In a truly free and liberal market anyone could hold fishing rights, sell any part of these to anyone else and buy more fishing rights. Dentists or PhD students could acquire fishing rights, as could investment fund managers. Companies could accumulate a large share of fishing rights, and actual fishers would have to lease or buy their quota from them. In the Danish VQS system, as in most others, there are restrictions on who can be a rights holder, as well as several limitations on transactions and concentration of VQS. Some of these are new to the VQS system, and others have been shaped over a long period of management and adapted according to the intentions and objectives behind the current program. These restrictions reaffirm the point made above that both the policy design and the market for fishing rights can be approached as social products. In the following section I will review the concrete elements in the VQS system.

---

[1] It should be noted that the VQS were given as private property to the operators with the condition that they could be revoked with an 8-year notice. It should also be noted that most people in the sector consider this scenario unrealistic under the current political setup. Reasons for this include general path dependency, the investments made in quota, its use as collateral, and the simple fact that few operators and their banks are likely to push for changes.

The Commodity and Principle of Catch History

An essential element in the establishment of a quota market was the creation of a commodity for rights holders to transfer between each other. The intention of the initial allocation was for management to distribute the quota shares to active fishers in the sector, which was done through the principle of historic catch. The principle of historic catch identified a vessel's share of the TAC during a reference period, which would then define its future allocations of fishing rights. Additionally, this share was transformed into a transferable commodity, as it could be sold or leased. In the case of the VQS system, this was organized by species and catch areas over a chosen reference period. The principle of historic catch thus has two main functions: it serves to set the individual allocation of rights (based on historical catches) as well as to include those who qualify and, consequently, exclude those who were not active during the reference period. This is how the initial distribution problem is solved. The difficult matter is then to choose a period of history. In the establishment of the VQS system, 3 years (2003, 2004, and 2005) were chosen but given different weight (20, 30, and 50%). The different weight of the 3 years reflects the intention to allocate the quota shares to active fishers. Once allocated, the VQS represents future fishing opportunities, as kilos of catch will be allocated annually based on the size of the individual vessel's quota share. Depending on the actual size of the TAC, allocations will fluctuate from year to year, representing a constant share of the changing TAC. The dynamics of changes in the annual allocations and their social implications will be explored in the following chapter, as I look more closely at the possibility to speculate on these fluctuations.

Establishing and designing the catch history is of course a critical element of this system. Not only does it exclude potential and historical users from the resource, as the VQS reflects a *vessel's* history, it also excluded a large group of active fishers at the time who did not own vessels. The historical catch is calculated from the vessel as a production unit as a whole, but it becomes the property of the vessel owner(s) only. This excludes the crew from their share in the catch history and consequently alters the relation between owner and crew. Both in principle and in practice, the vessel owners can sell the VQS without the consent of their crew, who were an important part of the historical production that granted the vessel owner these rights. The principles of common, free, and equal access—which since the 1970s had increasingly been managed through licenses and ration systems—were finally, in 2007, transformed into the exclusive private ownership of fish resources by vessel owners. As explained above, in one community a quota guild was established as a safeguard against this exclusion of the crew and share fishers, but this did not change the fact that crew and share fishers were not gifted quota in the initial allocation. The crew fishers were however, in this sole case, allowed to be part of the joint investment following the introduction. The catch history principle was used to create VQS as a commodity that could be transferred through market mechanisms by vessel owners. Since VQS can only be held by vessel owners, the rules regulating who can own a fishing vessel in fact also regulate the ownership structure of the VQS system: the so-called rights holders[2].

The Rights Holder

The rights holder is a person, company, or group of people with exclusive ownership of fishing rights and therefore the right to the stream of benefits derived from the resource. The definition of the rights holder thus determines who can hold and benefit from a fishing right. If fishing rights, for example, could only be held by a group structure, they would create a certain social formation and also condition the possible transactions that would have to occur between groups. If the rights holder can be an individual or alternatively joint-stock companies, then different social situations form. If the stated objective of fisheries policy is to keep fishing rights in the hands of fishers, it becomes critical to define the term "fisher." If this is done, for example, through the ownership of a fishing vessel, which then equates a vessel owner with a fisher, or through documented experience in the sector, so as to include crew, it results in two different scenarios. The definition of a rights holder is thus important, as it defines who can hold fishing rights, a factor that is hard to fundamentally change later, at least in regard to the initial allocation. It therefore also affects the dynamics of following market transactions.

In the political agreement of the New Regulation, a framework for the principles of the rights holders was established:

> It will be possible to merge vessels and quota shares. In case of merging the fish shall follow the tonnage. There will be established an upper limit for the amount of vessels, each individual fisher can have significant interest in. (Ministry of Food and Agriculture 2005, p. 1, author's translation)[3]

The New Regulation was followed by a concrete policy that linked the VQS to vessels (described above as "follow the tonnage"—that is, linked to the vessel), and thus established vessel owners as rights holders. This means that only vessel owners and company structures that own vessels can hold fishing rights. The question therefore arises, who can own a fishing vessel? Legally, a fishing vessel can be owned in Denmark by either a commercial fisher with "A" status or by a company of which two-thirds is held by commercial fishers with "A" status. A commercial fisher with "A" status is defined as a person who: (1) has Danish citizenship or who has been living in Denmark for at least 2 years, (2) has been occupied as a commercial fisher with "B" status (having an income from commercial fishing) for at least 12 months, and (3) has at least 60% of his income from commercial fishing. A potential vessel owner therefore must live in Denmark, have at least 12 months

---

[2] To my knowledge the term rights holder is a rhetorical way to avoid talking about owners. See previous note.

[3] "Fish shall follow the tonnage." It is interesting to note that one of the first public texts on the New Regulation did not talk about "access" or "fishing rights" but already had the "fish" phrase established.

experience as a commercial fisher (implied in the "B" status), and be dependent on income from fishing. This is the technical way to ensure the political objective that vessels are owned by the active fishers. Although the intention of these rules is to keep the fishing rights in the hands of owner operated vessels; the regulations technically also allow people outside the fishing sector to be financially involved by having 33 % ownership in fishing companies. Technically the same person could have several 33 % ownership scenarios in different operations. If we include the financing of vessels through loans, it blurs the ownership situation a bit more. Of course, in this way the bank sector indirectly owns a large share of the fishing rights in Denmark. But in addition to this, with holding companies and similar private financial arrangements, someone not directly active in fishing can receive income from fishing and from the leasing out of the VQS held by a vessel. However, while there is a clear intention to keep ownership in the hands of active vessel owners, the above definitions of "vessel owner"—and therefore rights holder—opens up a plurality of ways to organize a fishing operation, which will be further examined in the following chapters, particularly in Chap. 5, "Access and Fishing Activities."

Transactions

Transactions are transfers of fishing rights between rights holders. Setting limitations on transfers will determine the flow of fishing rights. These limitations can serve to keep fishing rights in the hands of certain groups or in certain regions, a measure commonly known as safeguarding. As most fishing fleets consist of several segments, transfers can be limited to occur inside each segment (i.e., small trawlers could only undertake transactions with other small trawlers). The segmentation could technically be achieved on the basis of different factors such as target species, gear type, geography, and vessel size. It is most common to use vessel length as a segmentation tool. For example, vessels over 12 m in length would not be able to buy fishing rights from vessels under 12 m and vice versa. Segmentation by gear type would result in another set of segments, and so forth. The limitations put onto market transactions potentially have a significant influence on the distribution of fishing rights and the development of the sector. In the following section, I will explain the different policy features in the VQS system relevant for the transfers of quota shares. Central to this system is a regulation to avoid excessive concentration of quota to single companies and measures to safeguard certain fleet segments.

Concentration and Maximum Ownership

There are several reasons for the application of a limit to maximum ownership in a market-based fisheries management system. First of all, there could be a political incentive to avoid too much contraction in the industry and to keep a certain number of operators and jobs in the sector, thus indirectly spreading employment and wealth. Second, it would make sense to avoid monopolies, where a few vessels could influence and manipulate the price for leasing and trading of fishing rights. The regulations therefore aim to avoid a situation with either too few buyers or too few sellers. Third, a high concentration of fishing opportunities can increase the dependency for others on leasing as a distinct income strategy, which was considered morally corrupt in the sector. In Denmark, this problem was solved by setting a maximum number on the vessels one person or company could acquire, therefore limiting the transfer of quota to their original vessel. In coordination with the industry, four acquisitions per individual were set as the maximum. In practice, this rule meant that in addition to the original vessel, one could only acquire fishing rights from four other vessels and transfer those rights to the original vessel. The double condition, *acquire and transfer,* is important to note, as it became important later. The anti-concentration rule was directed towards regulating the reduction of vessel numbers and not the size of quota shares, although the limitation in principle sets a technical upper limit to the quota share (which would be the five largest shares transferred to the same vessel). The intention was to allow the sector to restructure and for vessel owners to sell, while not allowing the largest operators to accumulate more than four vessels.

In April 2012, 5 years after the initial system began, the anti-concentration rules in Denmark were changed from the *four-vessel maximum* to individual maxi-

Table 3.1 New quota concentration rules were introduced in 2012, setting a maximum ownership of quota by percentage per person and vessel

Quota concentration rules 2012 (%)	
Cod in the North Sea	5
Cod in Skagerrak	5
Cod in Kattegat	5
Cod in the Eastern Baltic	10
Cod in the Western Baltic	5
Plaice in the North Sea	6
Plaice in Skagerrak	7.5
Plaice in Kattegat	7.5
Plaice in the Baltic Sea	5
Pollack/coalfish all catch areas	10
Haddock in the North Sea	10
Haddock in Skagerrak and Kattegat	10
Norwegian lobster in Skagerrak, Kattegat of the Baltic Sea	10
Norwegian lobster in the North Sea (EU zone)	10
Norwegian lobster in the North Sea (Norwegian zone)	10
Monkfish in the North Sea (Norwegian zone)	10
Sole in the North Sea	10
Sole in Skagerrak, Kattegat, and the Baltic Sea	5

mum percentages of ownership (of specific species in specific catch areas) (see Table 3.1). This change was implemented in recognition of the increasing evidence that the four-vessel limit had been a failure and was becoming inadequate as a management tool. Members of the leasing pools could easily see this inadequacy because identical contact details were listed in relation to more than 20 vessels. This was due to several reasons: low compliance by the rights holders, low control from the authority, and perhaps an unclear starting point.

In the course of a few years, a whole range of technical desktop solutions had been established to overcome this rule. Most importantly, the phenomenon of *acquiring and leasing,* instead of *acquiring and transferring,* made the four-vessel limit totally inadequate. Under this method, more than four vessels could be acquired and the quota leased back to oneself, in contrast to transferring it permanently onto a single vessel. In real terms it leads to concentration, though in legal and management terms the method complies with the rules. Others chose not to inform the authorities of the number of acquisitions they had undertaken—literally simply leaving that specific field in the form empty (personal conversation). That the four-vessel principle was a good idea in theory, but unclear and difficult to implement in practice, is demonstrated by a rather extreme example quoted here from an Irish magazine discussing the Danish VQS system:

> The HM45 was bought as new for 40,000 DKR or € 5,300 in September 2007 by a fisheries company called August A/S (August Ltd). August A/S sold the vessel in January 22, 2008, to a consortium of three people, one of whom was Tamme Bolt, the managing owner of August A/S. The price was now nine times higher at 370,000 DRK (€ 49,300). Three weeks later, on February 13, 2008, the HM45 was bought back into August Ltd. The price had now reached 32,370,000 DKR, or more than 800 times its value just five months earlier. (O'Riordan 2012, online document)[4]

The strange behavior and dramatic increase in price can only be explained by the value of the quota linked to the vessel. When HM45 was bought back by the same company who sold it 3 weeks earlier, HM45 suddenly held 7% of the total Danish quota for plaice in the North Sea, plus almost 10% of plaice in the Skagerrak. This is a very high amount, especially considering that the HM45 is a fiberglass vessel under 5 m in length, with no engine and no wheelhouse. HM45's function was likely to be a quota holding vessel in a transaction of quota from one company to another. The problem is, if the quota transfer should be regulated according to the four-vessel rule, should HM45 count as the multiple vessels that have had their quota transferred, or would HM45 just count as one acquisition for August A/S? This example is even more complicated than it seems at first, as one person appears to be involved on both sides of the transfer. Clearly catch histories from several vessels have been allocated to the boat, but for the company itself this represents only one acquisition.

With a low material value, low maintenance costs, and the probability that it is stored in a garage or garden with no moorage fee, HM45 is a prime example of the

---

[4] http://www.inshore-ireland.com/index.php?option=com_content&task=view&id=1010&Itemid =154 (Accessed June 01, 2012).

creation of flexibility in the regulation and a case where the vessel, as a material and cultural part of fishing practice, has become a pure adjunct to the quota market economy. To be more precise, it seems that the vessel is attached to the quota as is the paperwork, and not the other way around. Because the policy required vessel and VQS to be linked for at least 2 years after the acquisition, a large number of vessels were still in the harbors and referred to as "quota vessels" or "ghost ships." HM45 is just one of the five small vessels of the exact same kind within the August A/S company; and none of them take part in actual fishing, which is done from six other vessels. Through the leasing system, each year August A/S can move VQS from the quota holding skiffs to the actual active vessels, without technically transferring them and violating the four-vessel rule.

August A/S and HM45 are perhaps extreme examples, but they are not the only ones. We can only speculate that the public authority has been too focused on reducing the number of operators in the sector (and thus interpreted the policy with the highest amount of flexibility), or that the legal text is too weak and consequently power has been taken out of the hands of the authorities. In the words of Mogens Schou, the adviser to the Danish fisheries minister who had a significant impact on the policy, flexibility was very important in the design:

> A flexible use of the quotas was given high priority. The ITQ model allows for both structural adaptation through permanent selling of the shares and for day-to-day flexibility by allowing leasing of quotas and co-operation in fish pools. Also, the pool system was aimed at supporting co-operation in the individual harbours on the best use of the available fish and on the use and development of the harbour infrastructure and facilities. (Schou 2010, p. 19)

It seems retrospectively that flexibility was not only a high priority, but also a difficult policy attribute to control and manage. From the text qualifying the new 2012 anti-concentration rules, we learn that in the old regulation "some fishers are *forced* to have quota shares on vessels that are never actively fishing" (Ministry of Food Agriculture and Fisheries 2011, p. 1, author's translation and emphasis). It further states that the four-vessel rule does not prevent vessel owners from owning more than four vessels, only that the VQS cannot be transferred between these vessels, which in turn forces fishers to have inactive boats in the harbor (or in the garden). The framing in the policy argumentation leaves an impression that something is being swept under the carpet, or at least not being evaluated against the initial political intentions. If the intention of the four-vessel rule was to avoid concentration, this must be done through a clear limit on maximum ownership. Giving the quota shares to the vessels and basing the anti-concentration rules on numbers of vessels made the management of the fishery more complicated and was contradicted by the flexibility in the leasing system. It is unclear at the time of writing how the new anti-concentration rules will function. Compared to other systems—for example, the Alaskan halibut, which have maximum limits between 1 and 1.5%—the maximum limits in the VQS system between 5 and 10% are relatively high.

Another critical question is whether the interpretation of leasing and transferring will remain the same. If so, we can imagine a situation where a vessel owner owns the maximum permissible amount and on top of that leases from other companies, or vessels in the same company. For example, three individuals could each own a third of a company with three vessels (one active and two in the garage). Will they be able to have the maximum quota share on each vessel but annually lease from the two inactive vessels to the active one? This would mean an actual catch three times higher than the maximum concentration limits. According to the administration, the new rules only apply to permanent transfers (personal conversation). To avoid this situation, a limit on a vessel's total *activity* would have to be introduced, but this is not the case. At the end of the day, the system was designed to improve the economy of the fishery by reducing the fleet:

> While it is perfectly possible to ensure a structure benefitting small-scale fisheries, it is not within the logic of ITQs to allow overcapacity to persist. Thus, introducing ITQs in a situation with overcapacity will result in fewer vessels and empty spaces in some harbours. (Schou 2010, p. 21)

The central idea is that those fishers who are better and more efficient at fishing (in an economic sense) will buy up the rights from those who are not as efficient. But has this been the case in practice? In the following section, I will look at the developments in average sizes of quota shares since 2007.

Concentration and Average VQS Size

The following analysis is based on official data from the Ministry of Food, Agriculture, and Fisheries (VQS sizes and catch registrations in 2007 and 2011); the EU fleet register (for identifying gear types in 2007 and 2011); and some corrections to this data based on my personal observations and interviews. The Danish demersal fishery is divided into five catch areas: the North Sea, Skagerrak, Kattegat, Western, and the Eastern Baltic. Each VQS gives the right to land a share of the TAC of a species in one of the five catch areas. The following analysis thus has to treat each catch area and species individually. To illustrate the developments I use cod VQS, as cod is considered the most important demersal species economically in Denmark and has been the driver of much of the investments made in quota between 2005 and 2011. As we will learn in the next chapter, trade between vessels began before the system was introduced, which means that 2007 might not be the best basis for understanding the transformation. It is however the first year with VQS data. The data thus covers a period from the formal beginning of the system to the last year before the changes were made to the anti-concentration rules. Tables 3.2, 3.3, 3.4, 3.5, and 3.6 show the average size of quota shares on vessels, with a focus on the two main gear types, trawl and gillnet. Each catch area consists of approximately 1000 VQS, and the average size of VQS held by trawlers has increased between 40 and 400%, while the number of operators has been reduced. Gillnetters have increased their average VQS size between 20 and 40%, while the number of operators has also been reduced.

Table 3.2 Average cod VQS size North Sea per vessel

North Sea	Gillnet		Trawl	
	Average VQS size	Number of operators	Average VQS size	Number of operators
2007	2.25	275	1.6	169
2011	3.3	159	6.3	100

VQS vessel quota share

Table 3.3 Average cod VQS size Skagerrak per vessel

Skagerrak	Gillnet		Trawl	
	Average VQS size	Number of operators	Average VQS size	Number of operators
2007	2.4	211	1.9	204
2011	3.3	128	3.1	153

VQS vessel quota share

Table 3.4 Average cod VQS size Kattegat per vessel

Kattegat	Gillnet		Trawl	
	Average VQS size	Number of operators	Average VQS size	Number of operators
2007	1.7	168	3.2	187
2011	2.4	112	4.5	144

VQS vessel quota share

Table 3.5 Average cod VQS size Western Baltic per vessel

Western Baltic	Gillnet		Trawl	
	Average VQS size	Number of operators	Average VQS size	Number of operators
2007	1.5	267	1.9	216
2011	1.8	202	3.7	143

VQS vessel quota share

Table 3.6 Average cod VQS size Eastern Baltic per vessel

Eastern Baltic	Gillnet		Trawl	
	Average VQS size	Number of operators	Average VQS size	Number of operators
2007	3.9	88	3.9	115
2011	5	82	7.8	93

VQS vessel quota share

According to Table 3.2, in the North Sea the average VQS size held by trawlers is now larger than the average size held by gillnetters, which seems to indicate a shift in the type of production. VQS from more than a 100 gillnetters have been transferred to trawlers, now holding twice as much VQS on average. In general the numbers indicate a concentration and consolidation of fishing rights onto trawlers, while also showing a less significant increase in the average size of VQS on gillnetters. The Eastern Baltic (Table 3.6) seems to be the area that shows the highest degree of this type of concentration on trawlers. Here 10 vessels catch more than 50% of the cod, and 25 vessels account for 75% of the catch. While one company holds 8% of the quota, the median share is 0.24%. In general, quota and catch are concentrated in fewer and larger units in fewer places.

Fleet Segmentation and the Coastal Fishery Safeguard

The above data and trends indicate an increase in the average size of quota shares, especially in the trawl fleet, but do not give an indication of trends in the size of the vessels. Market-based fisheries management can be designed so the fleet is segmented by size. Transfer of fishing rights is possible inside each segment but not between the segments. This type of segmentation is often called a safeguard, as it serves to safeguard the rights of certain "groups" from being bought up by other "groups." The basic feature in fleet segmentation is that fishing rights can only be transferred between right holders inside each segment, often with the exception that fishing rights can be transferred from one segment to another but not the other way around, thus strengthening that specific segment.

Fleet segmentation in the VQS system is more complicated. On initial inspection, it appears to be somewhat similar to the above description. There is a certain segment for "coastal" vessels, which are defined as vessels less than 17 m in length with at least 80% of their fishing trips less than 72 h in duration. This technical definition was made to underpin the political objective of safeguarding the coastal fishery. As such it is an attempt to grasp the qualities of the heterogeneous fleet of smaller operators that mainly fish from one port and most often return home every day. Each year, members of this coastal fleet segment receive extra allocations of cod and sole on top of their annual allocations. The VQS in the coastal segment can only be transferred to other vessels inside the same segment, while VQS from outside can be transferred into the coastal segment. This is one way that the transfer system aims to ensure that the coastal fleet segment will maintain its size or even grow. However, as I will show below, this is not the case in practice.

The coastal safeguard measure was made optional, and vessel owners had to sign up for a period of 3 years at a time. The optional element of the safeguard measure was based on the wishes of fishers themselves,[5] but it created a weakness

---

[5] "Please do not force us into a box" was the wish, according to the Danish Fishermen's Organization.

in the design of the objective. First, since prices are higher for vessels and quota shares outside the coastal safeguard (where financial power and demand in general are greater), this came in practice to mean that any skipper with thoughts of selling his ship would not join or prolong his membership to the coastal fleet segment (personal conversation). Thus, when the coastal vessels choose not to prolong their membership to the coastal fleet segment, they automatically "migrate" to the larger segment and receive a higher price if they choose to sell. This created the possibility that a vessel owner could accumulate fishing rights inside the coastal fleet segment and then later "migrate" to the larger segment. In other words, here is a rather clear contradiction between the political objectives of the system (keeping jobs, value chains, and activities in the coastal fishery) and the priorities of the individual maximizing profit on the quota market. The moment a vessel owner plans to sell, he or she can leave the safeguard measure and use the market to maximize the income from the sale of his or her vessel. This strategy is common, so much so that there is a lack of both sellers and buyers within the coastal safeguard measure. In addition, the size of the extra allocations has been criticized. The steering board with coastal fishers overlooking the coastal safeguard made an evaluation in 2009 and concluded:

> The Coastal Fisheries Committee firmly believes that the coastal fishing scheme was made with the intent to preserve and develop the Danish coastal fisheries, which the current scheme does not seem to help. Based on the analysis conducted and based on many conversations with coastal fishermen around the country the Coastal Fisheries Committee assesses that coastal fisheries in Denmark are virtually being phased out. (Living Sea Denmark 2012, [2009], p. 3)

One of the problems identified in the report was the distribution of the extra allocations, which were distributed relative to the VQS size already held by the vessel owners. Thus, vessels with large shares receive a larger share of the extra allocations. This gives a rather unequal distribution, which is shown in the evaluation report:

> 7 vessels in the North Sea have been awarded extra 0–10 kg cod in 2009.
> 18 vessels in the North Sea have been awarded 11–100 kg extra cod in 2009.
> 43 vessels in the North Sea have been awarded the 101–500 kg extra cod in 2009.
> 27 vessels in the North Sea have been awarded the 501–1000 kg extra cod in 2009.
> 22 vessels in the North Sea have been awarded the 1001–2000 kg extra cod in 2009.
> 1 vessel in the North Sea has been awarded the 2001–5000 kg extra cod in 2009.
> 1 vessel in the North Sea has been awarded 5001–10000 kg extra cod in 2009.
> (Living Sea Denmark 2012, [2009], p. 3)

The problem seems to be that some vessels join the coastal safeguard system with large VQS shares and thus, because of the distribution principle, receive large portions of the extra allocations, which were intended to be allocated to small vessels:

> The Coastal Fisheries Committee anticipates that in the future new big players will come to the coastal fisheries segment, who will take as big amounts of the extra coastal fish that the scheme will be eroded. This will result in a large part of the vessels, for which the scheme was intended, in reality, getting nothing out of the scheme, and it seems clear that many smaller coastal vessels will not join in a new period under the currently applicable conditions. (Living Sea Denmark 2012, [2009], p. 3)

The phenomenon of the coastal fishers leaving the safeguard measure and selling their vessels illustrates the dynamic relation between being part of a group and, at the same time, an individual that can benefit from the higher VQS prices on the larger market. Since the program is optional, the individual has to make that decision. The evaluation report and my research indicate that currently maximizing the income from the sale is the preferred option. This can very well mean not giving youngsters or people from the same community a fair chance to buy the vessel and quota, meaning the fishing activity may disappear from the community. In order to secure a coastal segment in the long run, the implementation of a stronger design and "boxing in" of the coastal fleet could have limited the trade to "outsiders," but such changes are difficult to apply now. It should be noted that the extra allocation contributes to a better economy for each vessel owner in the coastal segment—or at least those who get the largest shares. The problem is that in the longer run operators will sell their VQS on the most lucrative market. Again, let us take a look at some of the statistics that point to some interesting phenomena.

Transfers Between Fleets Segments

As described above, a safeguard measure was put in place with the objective of securing the segment of so-called coastal fishers, defined as those with vessels less than 17 m long. These vessels could, if they obliged to a set of restrictions, sign up for a scheme giving them extra allocations of cod and sole quota, based on the size of their existing VQS. The scheme is optional, and vessels that take part cannot be sold outside the safeguard scheme without first leaving the scheme and losing the extra allocations. Tables 3.7 and 3.8 are based on all vessels, not only those in the coastal safeguard. Since 2007, the VQS share held by vessels less than 17 m long has decreased in three out of five catch areas. The data is partly skewed by "inactive" quota-holding vessels such as the HM45 mentioned above. Reductions in quota holdings are between 2 and 22 % (Table 3.7); but since vessels under 17 m can sell their fishing rights to vessels over 17 m, it can very well be expected that this decline will continue. This data is supported by interviews in which vessel owners explained the dynamic. If owners intend to sell, they leave the optional safeguard measure and can accept the better offers from larger companies. Some even doubt if there is a market for "coastal VQS" at all.

Table 3.7 Share of cod VQS shares held by vessels under 17 m long (out of +/- 1000)

Catch area	2007	2011	Change in %
North Sea	507.79	494.55	−2.6
Skagerrak	637.88	637.21	−0.1
Kattegat	660.1	587.79	−11
Western Baltic	602.04	663.3	+10.2
Eastern Baltic	699.15	544	−22.2

Table 3.8 Ownership and actual catch among less than 17-m long vessels. DØX is an administratively created catch area geographically similar to the Eastern Baltic, but with fish obtained by swapping from other countries

Catch area	2011 VQS registered on less than 17-m long vessels (%)	2011 Actual catch among less than 17-m long vessels (%)
North Sea	48	19
Skagerrak	61	58
Kattegat	57	65
Western Baltic	67	70
Eastern Baltic (not DØX)	53	45
Eastern Baltic (DØX)	67	57

VQS vessel quota share

Table 3.7 includes only cod VQS and the specific features of each catch area, and its fish have a significant influence on the development. Thus, the great reductions in the Eastern Baltic catch area might be explained by the high "catchability" or "fish per hour" in this area over the last few years. This makes large-scale trawling economically more attractive. Also, a regulation limiting days at sea in the Baltic areas contributes to the attractiveness of these quota shares, which give a high number of days at sea (160 in 2012). The opposite might very well also contribute to the explanation of the distribution of VQS in Skagerrak, where the cod has been "harder" to catch over the last few years. Dispersed fish makes trawling a poor business. However, the numbers do not indicate a radical shift in the distribution of small and large vessels. However, something interesting happens when the *actual catch* is included in the table instead of just the portions held by the vessels (Table 3.8).

Surprisingly, if the actual catch is integrated into the analysis (Table 3.8), we see that a number of vessels less than 17 m long must be leasing or swapping cod VQS to larger vessels, most notably in the North Sea where the figures drop from 48 % in ownership to 19 % in actual catch. There can be multiple reasons for this, which can be hard to identify from the numbers alone. Due to restrictions on transfers to one vessel from acquisitions, large companies can have vessels holding rights and then lease these from themselves, avoiding the former maximum transfer rules. As we have seen, small glass fiber skiffs are suitable for holding and semi-transferring of rights because of their low material value and maintenance. As mentioned above, one small glass fiber vessel in 2011 held substantially large shares of VQS but was not responsible for any catch. Such arrangements may very well distort the numbers. Another reason could be that smaller vessels can secure a stable income from leasing out their cod VQS (in times of high demand and good leasing prices) and instead focus on species where the relation between leasing and auction price is favorable to the (labor intensive) catcher, or on the few species still managed as a common quota. The examples show how complex the value chain is in a market-based fisheries management regime and that new economic strategies can exist around the phenomena of leasing in and out—a point that the numbers above do not reveal. Chapters 4 and 5 will examine this in much more depth.

Quota Blocks and Gear Differentiation

Less Active Vessels (LAVs)—vessels with an income from fishing in the reference years that was less than 224,000 DRK—were left out of the VQS system. Instead these operators were given equal allocations calculated from their combined historical catches during the reference period. However, the LAV licenses are transferable and thus give access to a share of the TAC. In addition, they cannot be bought up by vessels already holding VQS or a LAV license. This I call a quota block, because it is a block of quota that cannot be accumulated or broken up. On the other hand, through membership of a quota pool, LAV license holders can lease fish from VQS vessels and thereby supplement their annual shares (or quota block) with extra amounts. Although initially not considered a part of the VQS system—in fact excluded from it—the fact that LAVs can lease from the pools makes them part of the system's overall dynamics. It can be questioned if entering the industry, or perhaps more likely retiring as a LAV supplied with leasing, can be a viable strategy—as the LAV is more predictable in its annual fluctuations. Paradoxically, because the quota blocks cannot be accumulated in fewer units, the system for *less active* vessels will contribute to having more fishing *activities* spread out. LAVs are typically one-man operations and are a minor portion of the fishing sector. However, they represent an interesting alternative and supplement to the VQS system and perhaps could be a model for the future coastal safeguard.

The VQS system itself is multispecies and covers most commercial demersal species in Denmark. This means the system covers many fish species caught with a range of different gear types and fishing operations. Since the gear type used has implications for both the ecosystem and the social organization of production, it could make sense to differentiate between gear types. From a societal perspective, this would be a way to regulate the impact of the fishery on ecosystems. For example, if gillnets are considered more environmentally sustainable than other gears in a specific fishery, it could be an objective to safeguard and promote those operators using that specific gear type. On the other hand, it is a limitation on the individual choice and flexibility of fishers to choose freely between gear types.

The VQS system does not differentiate between gear types. As a result, quota shares can be transferred and leased between vessels with different gear types without limitations. Compared to the initial situation in 2007, trawlers now hold a larger share of the VQS (Table 3.9). Again, the difference between ownership and actual catch is significant. In almost every catch area, trawlers caught more than their combined ownership share of VQS would entitle them to. Some of this can of course be leased from their own quota-holding vessels, reflecting the fact that the trawlers in general have been buying VQS linked to gillnetters, although these are still registered as active due to technicalities in the regulation. Tables 3.9, 3.10, 3.11, 3.12, and 3.13 illustrate the changes in VQS ownership between gear types. For example, in the North Sea, vessels registered as trawlers were allocated 26% of the cod VQS in 2007, a share that has grown to 35% in 2011 (Table 3.9). However, the actual share caught by trawlers in 2011 was 48%. Since the actual catch reflects the

Table 3.9 Distribution and actual catch among gear types in the North Sea

Gear	2007 (initial allocation, %)	2011 (%)	Actual catch 2011 (%)
Otter trawl	26	35	48
Gillnet	60	50	32
Danish seine/flyshoot	8	8	18

Table 3.10 Distribution and actual catch among gear types in Skagerrak

Gear	2007 (initial allocation, %)	2011 (%)	Actual catch 2011 (%)
Otter trawl	38	46	44
Gillnet	49	41	40
Danish seine/flyshoot	8	7	14

Table 3.11 Distribution and actual catch among gear types in Kattegat

Gear	2007 (initial allocation, %)	2011 (%)	Actual catch 2011 (%)
Otter trawl	57	62	72
Gillnet	27	26	25
Danish seine/flyshoot	9	5	1

Table 3.12 Distribution and actual catch among gear types in the Western Baltic

Gear	2007 (initial allocation, %)	2011 (%)	Actual catch 2011 (%)
Otter trawl	40	54	58
Gillnet	40	36	36
Danish seine/flyshoot	8	4	4

Table 3.13 Distribution and actual catch among gear types in the Eastern Baltic (including DØX—an administrative "extra" catch area, which are fish in the same area as the Eastern Baltic that come from swapping with other countries)

Gear	2007 (initial allocation, %)	2011 (%)	2011 DØX (%)	Actual catch 2011 combined (%)
Otter trawl	43	69	57	85
Gillnet	37	30	38	15
Fictive vessels (Flying Dutchmen)	16	3	5	

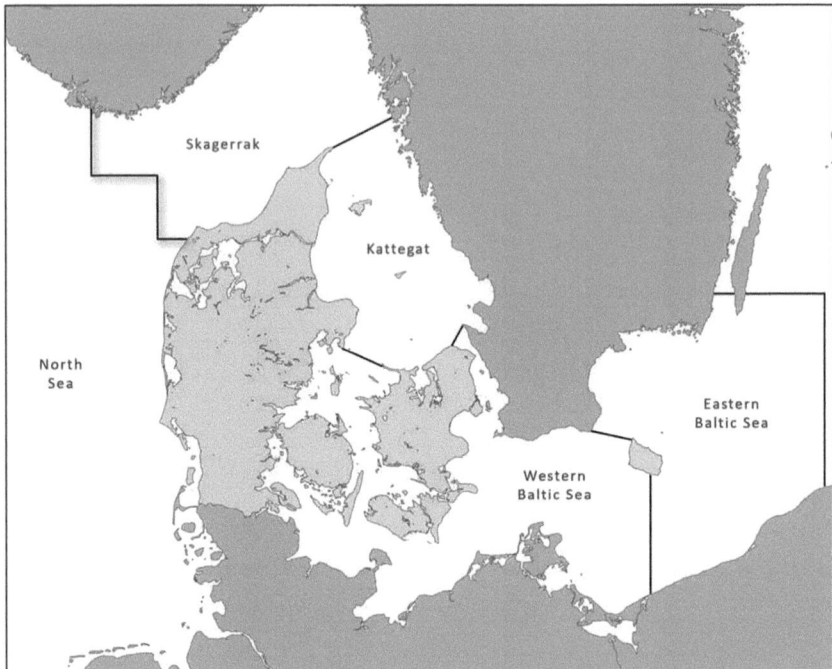

Fig. 3.1 Map showing the administrative defined catch areas in Denmark

activities at sea, they have also been included in the tables[6]. In some catch areas the change between the two are significant, while in others like Skagerrak, for example, there has only been a smaller shift between the two. The tendency however is clearly an increase in VQS to trawl vessels—both in ownership and in actual catches.

Geography: Area Specific Shares

The VQS is administratively split into five catch areas (see Fig. 3.1). To be able to fish in a catch area a vessel will need fishing rights for that species in that specific catch area, for example cod in the Eastern Baltic. This geographical division carries on from earlier management forms. As explained in Chap. 2, the minister was given

---

[6] This database is far from perfect, and in the data used to determine each vessel's gear type, I have observed a number of mistakes. The tendency is that vessels registered as gillnetters have changed gear without this information added to the database. Also in each area a portion of the VQS was given to nonexisting vessels, socalled Flying Dutchmen, which are papers holding rights but without a gear type registered nor any actual catch. Flyshooting is a method similar to Danish seine, but with the vessel moving forward while dragging in the seine. These are registered as Danish seine, but my knowledge of the vessels has led me to include flyshooting, which is not a category in the EU fleet register.

the privilege to divide and manage the fisheries both sectorally and geographically in 1979. The geographical division is mainly related to biological objectives and has nothing to do with the street address of the rights holder or restrictions on which port the fish have to be landed in. Examples of the latter are known from other countries (i.e., Norway), but this is not the case in Denmark. However, the area specific shares do have some implications for the strategies and everyday practices of the fishers. In Denmark, the special forms of the "days at sea" regulation in the Baltic Sea make quota shares here attractive. Ownership of quota shares here automatically gave the quota holder 160 days at sea in 2012—which in other areas would be considered a luxury. On the other hand, the high quality, size, and high price for cod in Skagerrak makes these quota shares the most valuable cod quota shares, in turn influencing the prices, demand, and supply. Since at least in the beginning quota shares were acquired by acquiring the entire vessel (including all fishing rights), the regulations mentioned above greatly impacted the trade and movement of the vessels and quota shares.

Fewer Boats and Empty Harbors

As has been discussed above, introducing market-based fisheries management will most likely result in empty space in some harbors. The numbers in Tables 3.2–3.6 also indicate a decline in the number of operators in each catch area. This is both a sign of vessels focusing their activities in fewer areas and people exiting the sector by selling their vessel and fishing rights. The map in Fig. 3.2 is generated from numbers of commercial fishing vessels (not only VQS vessels) per harbor. Many small harbors, along with some larger harbors, have disappeared since 2005 as active commercial fishing harbors, and many more have half or less of their 2005 number of vessels. Areas around Kattegat, the belts and fjords, have seen the greatest loss of commercial vessels, while the increase and growth in quota shares has been mainly along the Northern and Western coast of Jutland. Growth—measured in combined quota shares—has been concentrated in very few places and even those harbors have seen a decline in number of active vessels. A harbor like Nexø adjacent to the Eastern Baltic Sea, which has increased its quota share, has also seen over 20 vessels disappear. For the local economy this has several impacts, since the large trawlers do not always land their catch in Nexø but most often in Poland or Sweden. The local processing industry has thus further declined. Recently the two remaining fish buyers in the area merged their businesses and refused to pick up fish from the smaller harbors. At the other end of the country, in Northern Jutland, the four largest fishing harbors (Hirtshals, Skagen, Hanstholm, and Strandby) are all, or recently have been, enlarged. Longer piers and deeper basins are built in order to service larger vessels, not all of which are fishing vessels though. As a result, fishing activity is concentrated in fewer harbors. One feature of a market-based system is that fishing rights can be sold from one day to another. It is therefore difficult to invest in the attached value chain, since the value chain might very well go through

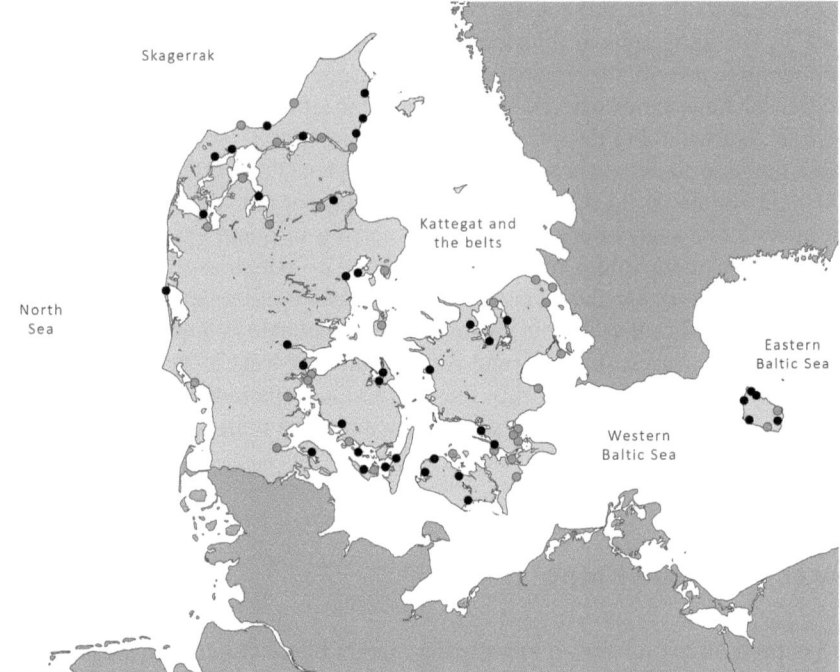

Fig. 3.2 Map showing decline in the number of harbors with commercial fishing vessels between 2005 and 2012. *Black dots* are harbors which no longer have any commercial fishing vessels, but had in 2005. *Grey dots* are harbors with half or less the number of vessels compared to 2005 numbers

other harbors or even other countries in the following year. As shown above and in Chap. 2, the introduction of the VQS system was mainly a reform intended to reduce capacity and improve economy for individual vessel owners. Both the number of people employed as well as the total number of vessels in the commercial sector (including vessels not part of the VQS system) has declined.

From 2005 to 2011, the average number of persons per vessel has fallen from 1.6 to 1.33, equaling a 32 % decline in people employed directly in the fisheries, while 18 % of the vessels disappeared. These numbers are for all commercial vessels including pelagic and other fisheries outside the VQS system. If we assume that as a condition each vessel will have at least one person on board as skipper, and consider the rest as crew, then the number of crew onboard has fallen by 50 %. Again the actual number is distorted downwards by the large number of vessels holding quota that are not actually involved in fishing. As a result of the many creative approaches to the quota trade, leasing activities, and the regulations governing them, the data sets today increasingly represent the quota as a paper asset, less the actual fishing practices on the ocean. It is in fact increasingly hard to clearly separate the two sides. A short example will illuminate this. In 2012 a national economic advisory council published a report on the status and wealth of the environment. The report estimated that the value of the fishery resource had dropped from 8 billion DKR in

1990 to 4 billion DKR in 2009. In other words, as a result of mismanagement the resource had lost 50% of its value (Flaaten 2013; Økonomisk Råd 2012). However, if the assumed profit rate was set at 5 instead of 2%, the result would be an increase in the resource value, now worth 10 billion DKR (instead of 4 billion). The report acknowledges that vessel values and profits are hard to calculate in a setting where the quota is embedded in vessel trading. It might be fruitful here to consider, like Brox does in a Norwegian context, whether the existing operators with their low level of capital and simple catching methods were not fully qualified to catch the available resource (Brox et al. 2006). There was, perhaps, no real need for further capitalization of the resource other than debt and economic difficulties in the sector. The idea of the *economic man* also guides the economic evaluations of the fleet performance, as much as they have guided the design of the VQS system. Large trawling units fishing large volumes might be profitable in one way, and thus represent fleet efficiency to the economist. However, quality and price differences are often left out of these nationally harmonized datasets, a concern that could also be raised in regards to environmental impacts of the increase in trawling. It could be asked if fleet efficiency and related fish products are desirable at all in the longer run, given the increased global competition on price (Vedsmand 1998).

Markets in Motion

The above analysis reveals that the policy design was more than just a technical management solution. Policy was formed as part of a political process, with a design to balance different objectives and interests. As such it is part of a process that is still ongoing, which the current discussion on coastal fisheries and quota concentration testifies. The anti-concentration rules were designed to avoid concentration. However, as I have shown, the policy was not managed appropriately by the administration and not in line with the political intentions. There is an apparent discrepancy between the rhetoric and the actual policy outcomes, which suggests that the policy is formed around the interests of the largest operators, allowing them to expand more than intended. As such this carries on from the political and organizational change described in the previous chapter, where larger operations were promoted and in time became the most influential members of the producer organizations. The administration, it seems, is itself unclear on the main objective of the new concentration rules. The responsible minister, Mette Gjerskov, announced that,

> With the new rules we have, popularly speaking, put a plug in the hole, so the quotas will not be concentrated any further. I see this as important in order to ensure wide activity in our fishing harbors in the future. (Ministry of Food and Agriculture 2012, press release, author's translation)

This indicates that we are dealing with a measure that ensures activity in as many harbors as possible (without limiting individual flexibility too much). As I have indicated above, there is little in the New Regulation that hinders further concentration. Currently there are only a few operators with such high degree of ownership,

Fig. 3.3 The port of Nexø has increased its share of cod quota but has still seen a vast number of vessels disappear. Despite the growth in allocations, fishing activities have concentrated on fewer operators. (Photo: Jeppe Høst)

and the new limits legitimize others to follow the same path. Worse, the flexibility via leasing still allows companies to exceed the limit described by the minister.

Market-Based Transformation

The VQS system transformed the *common access to marine resources into an individual commodity* that could be traded on a market. This was done through the principle of historic catch, which excluded not only past and future users but also altered the concept of equal access at the expense of the crew. Individual rights were given to vessel owners and made transferable. The distribution of fishing opportunities and activities is no longer the responsibility of the state but rather coordinated through market mechanisms. The parallels to Polanyi's account pose intriguing questions about the practice of market-based fisheries management. Was the privatization of fishing rights part of a capitalistic process, similar to the one where labor, land, and money were formed as commodities on a market? If so, what in the existing society was broken down and marginalized as adjuncts to the new market economy for fishing rights? The above analysis points at a transformation marked by ownership concentration and an increase in trawl fishing. Figures 3.3 and 3.4 show the concrete outcome, where even large harbors with large amounts of quota shares are evidence of the concentration on fewer operators.

Fig. 3.4 Quota shares were linked to the vessels and many vessels were acquired for the need of the quota only and therefore subsequently scrapped. (Photo: Jeppe Høst)

In this chapter, I have examined central elements governing the economic and social relations of fisheries and enabling changes in the relation between people on board: the geographical spread of fishing activities, the possibilities for the next generation, the course of future development, and so on. I argued that policy design would reflect a complex set of interests, and the enacting of the policy would reveal some of those conflicting interests as they were balanced by the state. In regard to the two most central societal objectives, avoiding vast concentration of fishing rights and safeguarding the coastal fisheries, the policy has not been able to deliver. In addition, the administration of the central policies has not been handled in accordance with the original political decisions nor evaluated and reacted to appropriately in relation to these. Unless outcomes for other species than cod are somehow notably different, this is the conclusion of the above assessment. Fishing rights—and to an even higher degree actual catches—have been concentrated on larger vessels registered as trawlers and ownership consolidated in larger shares. The policy, including the subsequent changes, therefore also enables and legitimizes large operators to accumulate to the maximum limit. As *one* boat can have the maximum limits in *all* catch areas, a plausible projection would be a sector dominated by 10–20 large units. This would be a rather significant change from the former distribution of fishing activities. These 10–20 vessels would be operated by the *captains of finance* referred to in the previous chapter, who have ventured into an expansive accumulation of fishing rights in the new market. In conclusion, the innovative designs highlighted by both the Environmental Defense Fund and the European Commission on Fisheries are not to be found. In fact, after careful analysis it is hard even to hint at where their conclusions could come from. Perhaps Denmark, with its reputation as a social democratic welfare state, was the perfect theoretical example of a market-based fisheries management system in social balance, but in practice it is not. As I have shown, the anti-concentration and coastal fisheries safeguard regulations have been inadequate and provided only small obstacles to market forces. The desired flexibility in the leasing system and optional nature of the safeguard regulations has proved a clear contradiction to the initial social objectives. In the following chapters, I will undertake deeper analyses of the qualitative sides of these changes both for large and small operators.

References

Aguilera-Klink, F. 1994. Some notes on the misuse of classic writings in economics on the subject of common property. *Ecological Economics* 9 (3): 221.
Arnason, R. 2002. *A review of international experiences with ITQs*. Portsmouth: Centre for the economics and management of aquatic resources, University of Portsmouth.
Beckett, Jens. 2009. The great transformation of embeddedness: Karl Polanyi and the new economic sociology. In *Market and society: the great transformation today*, eds. C. M. Hann, Keith Hart., and xi, p 320. Cambridge: Cambridge University Press.
Bonzon, K., K. McIlwain, C.K. Strauss, and T. Van Leuvan. 2010. Catch share design manual: A guide for managers and fishermen New York: Environmental defense fund.
Brox, Ottar, J. M. Bryden, and Robert Storey. 2006. *The political economy of rural development: Modernisation without centralisation?* Delft: Eburon.
European Commission on Fisheries. 2012. Transferable fishing concessions. Brussels: European Commission on Fisheries.
Flaaten, Ola. 2013. Institutional quality and catch performance of fishing nations. *Marine Policy* 38 (0):267–276. doi:http://dx.doi.org/10.1016/j.marpol.2012.06.002.
Gordon, H. Scott. 1954. The economic theory of a common-property resource: The fishery. *The Journal of Political Economy* 62 (2): 124–142.
Graeber, David. 2001. *Toward an anthropological theory of value: The false coin of our own dreams*. New York: Palgrave.
Grafton, R. Quentin. 1995. Rent capture in a rights-based fishery. *Journal of Environmental Economics and Management* 28 (1): 48–67. doi:http://dx.doi.org/10.1006/jeem.1995.1004.
Grafton, R. Quentin. 1996. Individual transferable quotas: Theory and practice. *Fisheries Reviews in Fish Biology and Fisheries* 6 (1): 5–20.
Grafton, R. Quentin. 1999. *Private property and economic efficiency: A study of a common-pool resource*. Dunedin: University of Otago.
Granovetter, Mark. 1985. Economic action and social structure: The problem of embeddedness. *American Journal of Sociology* 91 (3):481–510.
Gudeman, Stephen. 1985. *The anthropology of economy: Community, market, and culture*. Malden: Blackwell.
Gudeman, Stephen. 2008. *Economy's tension: The dialectics of community and market*. New York: Berghahn Books.
Hann, C. M., and Keith Hart. 2009. *Market and society: The great transformation today*. New York: Cambridge University Press.
Hardin, G. 1968. Tragedy of Commons. *Science* 162 (3859): 1243–&.
Hirschman, Albert O. 1977. *The passions and the interests: Political arguments for capitalism before its triumph*. Princeton: Princeton University Press.
Høst, Jeppe. 2012. Fairness or efficiency? *SAMUDRA* 61:15–17.
Living Sea Denmark. 2012. Evaluation of coastal fishing scheme, translated from Danish, ed. Kystfiskerudvalget. Denmark: Levende Hav.

Ministry of Food, Agriculture and Fisheries. 2011. *Indstillingsnotat vedrørende ændring af koncentrationsreglerne for IOK og FKA i Reguleringsbekendtgørelsen*. Copenhagen: NaturErhvervstyrelsen.
Ministry of Food and Agriculture. 2005. *New regulation*. Copenhagen: Ministry of Food and Agriculture.
Ministry of Food and Agriculture. 2012. *New concentration rules—press release*. Copenhagen: Ministry of Food and Agriculture.
O'riordan, Brian. 2012. Transferable Fishing Concessions. *Inshore Ireland* 8.
Polanyi, Karl. 1957. *The great transformation*. Boston: Beacon Press.
Ribot, Jesse C., and Nancy Lee Peluso. 2003. A Theory of Access. *Rural Sociology* 68 (2):153–181. doi:10.1111/j.1549-0831.2003.tb00133.x.
Rittenberg, Libby, Timothy D. Tregarthen, and Knowledge Flatworld. 2008. *Principles of microeconomics. Nyak*. New York: Flatworld Knowledge.
Sanchirico, James N., and Kailin Kroetz. 2010. Economic insights into the costs of design restrictions in ITQ programs Washington: Resources for the future.
Schou, Mogens. 2010. Sharing the wealth. *SAMUDRA* 55:18–23.
Smith, Adam, and Edwin Cannan. 2003. *The wealth of nations*. New York, N.Y.: Bantam Classic.
Townsend, Ralph, and James A. Wilson. 1987. An economic view of the tragedy of the commons. In *The question of the commons: The culture and ecology of communal resources*, eds. Bonnie J. McCay., and James M. Acheson. Tucson: University of Arizona Press.
Vedsmand, Tomas. 1998. Fiskeriets regulering og erhvervsudvikling - i et institutionelt perspektiv. Bornholms Forskningscenter, Nexø.
Økonomisk Råd. 2012. Økonomi og miljø:2012. http://p1kitapp01lcur.adm.ku.dk:8081/portal/da/publications/oekonomi-og-miljoe(0b8215d5-ef0b-40b4-9b4e-293543a60629).html. Accessed 10 Oct 2014.

Chapter 4
The Commodity and its Exchange

Abstract This chapter examines fishing quota as a commodity in both a conceptual perspective and through ethnographic examples. Inspired by Marx's ideas of the commodity, use-value, exchange-value, and ground rent, the chapter combines a theoretical approach with ethnographic material. The different aspects and the value of the quota are examined through the concrete exchange of fishing rights, and it is explained why quota trade can give rise to speculation and monopolies. In the final part of the chapter, it is argued that the value of transferable fishing quotas rely on a social relation between owners and nonowners of quota, as a form of monopoly rent.

Keywords Modes of operation · Large-scale fisheries · Small-scale fisheries · Social organization · Quota investment

The VQS Commodity

> All of a sudden by the end of the year when the value of the boat was estimated—it had increased two or three million in value, the quota, in one year. Mikael, the accountant, said that was something of a change. The banker came and shook my hand and congratulated me, 'congratulations with the title', I was one of those millionaires. (Personal conversation, December 2011, author's translation)

As we have seen in the previous chapters, the Vessel Quota Share (VQS) system was the establishment of a market for trading fishing rights, with those rights representing the opportunity to benefit from a share of the commercial fishery in Denmark. The policy was designed to balance objectives between individual and societal gain, between different individuals, groups, and communities in the fishing sector. The most difficult of these problematics was the balance between the desired concentration of VQS on fewer vessels and the spread of fishing jobs in coastal areas. The descriptive statistics I presented in Chap. 3 indicated that despite anti-concentration rules and a coastal fishery safeguard, fishing rights have concentrated among fewer rights holders and in fewer places. Furthermore, the statistics indicated that the trawl fleet has expanded in the 5-year period since the introduction of VQS. The analysis also revealed that quota leasing is much more widespread than can be explained by "a little flexibility" in regard to reducing bycatch. In this chapter, I turn to the commodity of VQS and through that to a qualitative examination of quota trade and leasing. By involving ethnographic material, I try to understand how the VQS as a commodity is related to fishing practices and how it is conceptualized in exchange situations. The economic literature tells us that:

> Transferability of the harvesting rights allows fishers with higher returns per unit of fish to increase their share of the harvest and, thus, should increase the rents from fishing. Consequently, with the adoption of ITQs there is a potential for increased profitability of vessels and the fishing industry. (Dupont et al. 2005, p. 32)

According to this theory, what we will encounter in the following analysis should be that the most efficient fisher buys rights from the least efficient, and quota leasing is only used as a matter of short-term flexibility. Needless to say, perhaps, the following examination will present material that is more diverse than this and to some degree contradictory to this theoretical proposal.

The Commodity as the Framework

To provide an analytical framework, I will start out by outlining the fishing right as a commodity, and then move on to describe the different situations generated by the exchange of it. With a greater conceptual understanding of VQS as a commodity and how it can support and generate different economic strategies, I will turn to the empirical findings from my fieldwork. The fishing right as a commodity is a multifaceted object: as "fish," it is largely treated as a commodity necessary for production. The fishing right is also an investment with certain risks and potential, in which case it appears as a financial asset. It is also part of nature and, like land, can be monopolized to provide a stream of benefits. Finally, the fishing right is a political and administrative management object that binds high-level political decisions with fishing activities. In interviews, fishers talk about these different aspects, often using the same notion, "fish," to describe it as both a means of production, management tool, investment, and actual product for the auction. The specific qualities of the VQS have implications for the exchange process, which the following analysis of the VQS as a commodity will illuminate.

It should perhaps be noted that this line of progression, moving from the commodity as a conceptual framework to the ethnographic material, is somewhat simplistic compared to the actual line of research that was undertaken. The discussion and analysis of the VQS commodity in the following sections is based on ethnographic material collected between 2010 and 2012, and on the synthesis and analysis of this material in the final year of writing and thinking. I found it interesting that the actors are related to each other through the marketplace. A buyer needs a seller, a leaser needs a rights holder to lease from, and they all need people to administer these transactions. Together it forms a network of dependencies and mutual relations. To comprehend this complexity, I found it useful to widen the understanding of the VQS as a commodity, and to outline the different qualities of this peculiar commodity that generate and frame the different ways to structure life in and around the VQS system. Therefore, this chapter begins with an analysis of the VQS as a commodity, outlining the different qualities that have implications for the way people conceive of the exchange of VQS.

The Commodity

In the first volume of "Capital," Karl Marx begins his inquiry into capitalism by looking at the commodity as one of the most basic elements in the capitalist system (Marx 1976). For Marx, the commodity "appears as its elementary form" and the capitalist mode of production appears as an "immense collection of commodities" (Marx 1976, p. 125). What Marx establishes is the commodity as the central object that relates people to one another. When we buy commodities, we are actually dealing with social relations of production: the things that were produced by people and with a specific distribution of the means of production. But, Marx argues, the focus on commodities hides and objectifies these social relations, obscuring the human aspect of production when we deal with commodities in everyday life. Therefore, the wording "appears as" in the phrase "appears as its elementary form" is important. We look at prices and see commodities, but we forget about the social relations and human labor behind the objects. The fishing right—in this case the VQS—is also to some extent a fictitious commodity (see Chap. 3) that is traded on a market and connects people in the fishing sector in different ways. Therefore, to examine the VQS, I will parallel Marx and ask what the fishing right as a commodity hides. What lies behind the commodity? What is structured by the commodity? What are the social relations inherent to the fishing rights as a commodity? What constitutes its value?

Use and Exchange Values

To understand the social relations of the commodity, Marx sets out to find the value of a commodity, which he finds in "socially necessary labor time." From this starting point, he explains the capitalist system, its dynamics, and the different classes that are present within it. As stated above, before Marx lays out the capitalist system, he starts out with a close examination of the commodity. For Marx, a commodity is an object or article that through its qualities can satisfy human needs (Marx 1976, p. 125). Likewise, the VQS commodity satisfies a human need, as it is a necessary factor of production for participants in the commercial fishery. However, in addition to being used to fulfill human needs, the VQS can be traded and exchanged. In Marx's words, a commodity can be viewed from two different perspectives: the commodity as a quality and commodity as a quantity (Marx 1976,

p. 125). These two views form the basis of a dialectic relation in the nature of commodities between a use-value and an exchange-value. The use-value is related to the commodity's capacity to fulfill human needs, and thus expresses the *quality* of a commodity. On the other hand, the exchange-value is related to *quantity*, as it is the expression of a commodity's value relative to other commodities[1].

> Exchange-value appears first of all as the quantitative relation, the proportion, in which use-values of one kind exchange for use-values of another kind. This relation changes constantly with time and place. (Marx 1976, p. 126)

In analyzing the VQS as a commodity inspired by Marx's work, we have to remember that VQS is not a genuine commodity: access to fishing was not produced as a commodity for a commodity market. In that respect, the VQS as a commodity does not share the most common feature of commodities: that they are products of human labor (Marx 1976). Despite this, VQS as a thing or article does satisfy a human need and materializes in the production of fish. However, even though VQS is commonly just talked about as "fish" or "quota," and a market has been established for trading VQS like other commodities, there are certain qualities in the VQS that deviate from a genuine commodity and that come from its *administrative production*.

The VQS and the VQS market are produced in a political and bureaucratic environment rather than by human labor. This production has influence on the commodity and subsequently on the analysis of the VQS as a commodity (which I will return to below). Despite these deviations, it is possible to find an exchange-value in the VQS as the VQS has a relative value in money (or equivalent VQS). It is slightly more complicated to point out the use-value in VQS. The VQS is a piece of paper, or rather a number on a piece of paper or in an immaterial database. So how does the VQS satisfy human needs? The VQS provides *access* to a fish resource that can be used in production: catching and landing of fish. The quality of the VQS is to provide access to this production. As a result, VQS is called a fishing right or harvest right, or even sometimes an access right. One kilo of VQS allocation provides access to harvest one kilo of fish. The use-value, put simply, is the access to a share of the commercial fish resource, and the opportunity to turn that fish resource into a fish commodity to be sold on the auction or to a fish buyer. As such, VQS is needed for the production and reproduction of a commercial fishing company. The exact use of the commodity—for example to produce canned fish—is left to the owner and confirms the heterogeneity of the use-value aspect. As mentioned above, I will deal with the use-value and how the access to quota is organized in the next chapter. For the following analysis of the exchange-value, we can therefore leave the use-value behind for later examination, but keep in mind that the VQS as a use-value commodity is a necessary means of production in the production of fish commodities (commodity this time defined as things created for exchange). It is necessary for access to the fish resource.

---

[1] Today the money commodity most often serves the purpose of measuring value and mediating trade; but, for example, in Iceland there is the phenomenon of the cod equivalent, used to measure fishing rights against each other. Also, states such as Denmark and Norway swap quota between each other without the mediating use of money.

The Two Aspects of the Rights Holder

Initially, it makes sense to talk about a use-value and an exchange-value in relation to the VQS commodity. But since the VQS is held by a vessel owner, we can prolong this point and examine two aspects of the VQS holder. We can call the two aspects the *use-value right holder* (concerned with the use of the VQS) and the *exchange-value right holder* (concerned with the exchange of VQS). Following the introduction of the VQS system, the two aspects have become part of the array of tasks a vessel owner has to be concerned with in his practice and career. However, the two aspects can also be separated, so that a practice is only concerned with the exchange-value or with the use-value, a point I will elaborate on later. One is concerned with the organization of use, while the other is concerned with the handling of exchange of VQS. By focusing on the two aspects, we can already see a clear contrast to the situation prior to the VQS system: an exchange-value has been added to the fish resource, and consequently a *VQS exchange-value practice* has been added to the vessel owner's practice. Before, fish for consumption markets were produced through the capital and labor invested in turning nature into a fish commodity; but now, ownership of VQS is a necessary condition for accessing commercial fish stocks. In other words, the VQS has become a part of the means of production, and after the introduction of VQS in 2007 the organization and trade of VQS has become a central aspect of fishing operations.

TAC and VQS

In order to fully unravel the implications of the exchange-value that was added through the introduction of the VQS system, we have to introduce the specific scientific and political context in which VQS is produced. Most important is the concept of total allowable catch (TAC), which is the total output of a given species, set annually. Since the VQS is a share of the TAC, the latter has significant impacts on both the exchange and use-value of the VQS. The TAC is set by the International Council for Explorations of the Sea (ICES) and the European Union in coordination. This takes place for most species once a year, most often in December. Based on calculations from catch samples, catch statistics, and virtual analysis of fish populations, the TAC represents a prediction of a dynamic natural environment. Since fish populations are never stable in space, size, and time, the TAC fluctuates from year to year. This is due to many factors—not only fishing pressure—including imperfect scientific models as well as political decisions, climate change, and pollution. The biologists in ICES deliver their recommendations, and the resulting TACs are dependent on the degree to which political decisions follow these recommendations. The TAC phenomenon is a pragmatic division of the scientific and political aspects of fisheries management. The result is that the tons of fish allocated to each state will change from year to year, and that the combined use-value (the kilos allocated to each VQS holder) will change from year to year as well. So while the share size of the VQS (i.e., 1%) does not change, the resulting use-value changes from year to year. This is a very peculiar quality of the VQS commodity, representing a constant share of a dynamic fish population.

The VQS is expressed as both a per-mille share of the TAC and as the number of kilos that this share yields in each year. The exact number of kilos—of use-values—allocated to a vessel is then in turn dependent on the size of the TAC that specific year. With a TAC at 100 t a 10‰, VQS ownership will result in a 1 ton allocation. In this way, the most basic features of the VQS system are a fixed share in per-mille and a changing annual allocation in kilos (see Fig. 4.1). The VQS holder will be impacted by the changes in the TAC, as his ability to reproduce the operation will be dependent on the size of the annual allocation: the quantity of use-values. If the TAC is set to zero—to use an extreme but not unrealistic example—the VQS holder will have to rely on other options to ensure a flow of production. These options could include leasing or investing in other species or simply downsizing production in order to stay in the harbor.

Fluctuations and New Opportunities

On the other hand, such fluctuations open up new ways to build economic strategies around the VQS. The fluctuations make it possible to invest in VQS when the exchange-value is low and to sell when the exchange-value is higher. As we will see from my interviews with rights holders, most VQS holders are aware of this aspect of the system and follow the development of prices for VQS closely. For those who are either expanding or selling their companies (and VQS), the fluctuations become even more critical. Timing of investments and sales can depend greatly on the developments in the exchange-value and the negotiations between ICES and EU on the horizon. The use-value and exchange-value in the following years is uncertain; and as a result, some of those VQS holders choose to sell their VQS and fishing operations, even though they would like to continue fishing for a few more years. But the two aspects, buying and selling at favorable prices, can also be combined into one distinct strategy as a quota speculator.

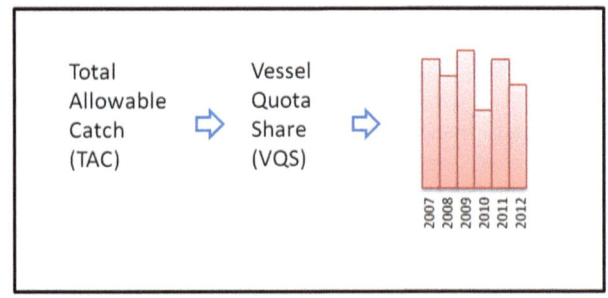

Fig. 4.1 Changes in TAC lead to fluctuating annual allocations while the VQS as a percentage however remains constant

Speculating in VQS consists of buying and selling VQS in order to benefit from fluctuating prices. In practice, this means buying at low exchange-value and selling again at higher exchange-value. Trying to foresee the developments in quota prices is central to this strategy, but in its purest form it is solely dependent on fluctuations in VQS. As the number of considerations and analyses involved in buying and selling grow, the process is more appropriately termed *investment* and to a lesser degree *speculation*. To use the same example as above, when the TAC is set to zero a quota speculator would use the opportunity to buy. Assisted by the economic problems for VQS holders dependent on that specific species, the speculator would be able to accumulate at a lower cost, knowing that the TAC would most probably be higher sometime in the future.

In principle, speculation as a distinct strategy requires the VQS holder to be economically independent from the income from use-value until the VQS is sold again. Of course, the VQS can be leased out until it is the right time to sell, which eases up the economic situation. Interestingly, for the speculator it is not so important what the value of the VQS is in relation to the real production prices and incomes. What is important is that the value fluctuates. Knowing this, it might be possible to comprehend why prices for quota shares can rise far higher than the income of production. This can be seen for example in Iceland, where during some periods a kilo of cod in quota shares was many times more costly than the landed value of one kilo (Einarsson 2011; Højrup and Schriewer 2012); or in Denmark, where prices for some species are 10–12 times the landed value. This speculative aspect forms the basis for the financialization of the fish resource. It is not unlikely in this perspective that speculation can drive prices up, which would contradict the economic theory mentioned above, as the market prices do not reflect the relative efficiency of fishing operations.

Exchange and the Marketplace

Exchange-value—the value of the VQS relative to other commodities—appears to be related to the expected use-value across a series of years. Often fishers measure it as the number of years that a quota has to be fished in order for the cost of the VQS to be paid off. The financial climate and accessibility of financial capital (i.e., interest rates) also seem to influence the exchange-value, in addition to demand for and supply of quota shares. On top of this are important factors such as the prices of the landed fish and costs of production such as oil. In the final part of this chapter, I will return to a theoretical discussion on the value of VQS and other fishing rights. The market—or the transaction through market mechanisms—is the last phase for this analysis of the exchange-value dimension of the VQS as a commodity. On the pages of the fishers' magazine, at the ship brokers or simply through face-to-face interactions, prices and transactions are negotiated and coordinated. All these and more constitute the marketplace for exchange of VQS. Here, VQS is sold and passed on to a new holder, and the cycle starts over. The industry-wide results of these transactions over time were examined in the previous chapter. As I have explained above, most VQS holders will always have one eye on market transactions. What is happening on the VQS market is discussed just as much as the TAC, water temperatures, and wind directions. But the permanent selling and buying of VQS is not the only way that VQS is transferred between rights holders. On an annual basis, VQS can be leased through quota pools. The dynamics involved in leasing are slightly different from those of permanent VQS transfers, and therefore should be dealt with separately.

Leasing

The dynamics described above were associated with a permanent transfer of VQS, and the price paid for the VQS appeared to be equal to several years of expected use-value: income from several years of material production based on that VQS and the resulting annual allocations. In contrast to this, leasing is the transfer of VQS for 1 year only. In principle, both the exchange-value and use-value are transferred, but only for the remaining part of that year. The leaser can lease the VQS to a third person, and thus speculate on the fluctuations in the exchange-value on the leasing market. Due to the short timeframe, this is more risky though not impossible. The cost of leasing—rent we could call it—seems to be more sensible to fluctuations in supply and demand. This is evident in November and December when large amounts of VQS for some species are put on the leasing market, whereas others are in short supply. As what is leased out is the VQS of a specific year—in the shape of the annual allocation in kilos—the leasing market is influenced by the dynamics of the annual cycle. This is another peculiarity of the VQS commodity: when it appears as an annual allocation in kilos, it has to be used before the 31st of December. This aspect of the VQS commodity is particularly relevant for understanding leasing activities.

Annual allocations not used at the end of the year are of no value. Therefore, VQS holders use the leasing system to earn an income from the excess kilos of their annual allocations. In principle, a precondition for the leasing system is that there is a limited amount of VQS—or in other words that the VQS is an exclusive commodity (like land)—and that some individuals have excess VQS-kilos and others a deficit. In other words, the lease system is a social relation between owners and nonowners. The specific properties of the VQS commodity have influence on leasing as a phenomenon. The exclusive ownership of a limited natural resource, or at least ownership of access to this resource, means that the leasing market is exposed to domination and unequal power relations. Those who were not gifted free quota in the initial allocation, and those with little or no VQS, are in a particularly difficult situation. I will further illustrate this process below through ethnographic material and discuss it in the last section on the value of VQS.

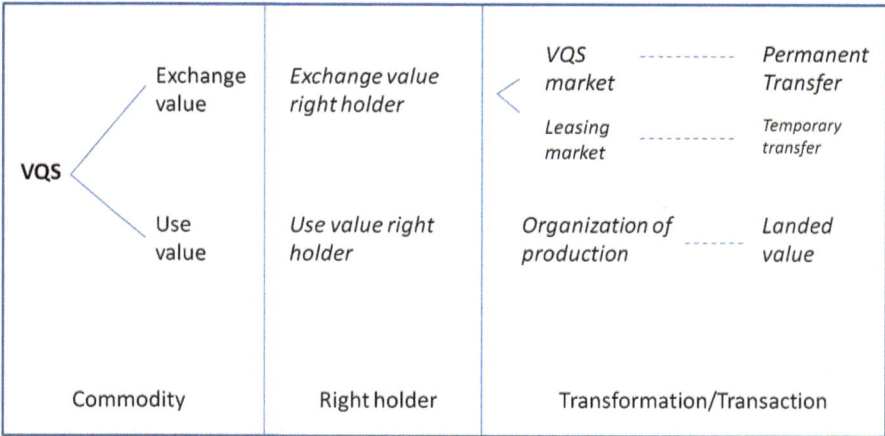

Fig. 4.2 Schematic illustration of use-value and exchange-value of VQS. With the introduction of transferable quotas an exchange-value was added to the fish quota and subsequently became an intrinsic part of the lives of quota owners and leasers

Exchange Situations

Permanent transfer and leasing of VQS can be combined into one economic strategy, where VQS is bought only to be leased out. For some—those with easier access to financial capital—buying VQS to lease out is a sensible strategy when profits made from leasing are higher than financial costs such as bank interests on loans. This is the basic principle of the economic strategy of those that have in Denmark popularly been called "sea-lords" or "quota-barons," people who gain from leasing fishing rights to "tenants." So far I have examined the VQS commodity theoretically and explored the exchange-value aspect in detail, while reserving the use-value aspect for the next chapter. The model in Fig. 4.2 shows the two different aspects of the VQS commodity in a schematic way. In the left box are the two aspects of the VQS as a commodity. These form the basis for the two rights holder aspects, shown in the middle box. On the right are the different actions or transformations that the rights holder can engage in. On the market, the VQS is either sold permanently or leased out as a temporary annual right. As a use-value, it is used as a means of production to create the fish commodity that can be sold on a commodity market. The model shows three analytical stages from commodity, rights holder, and finally the transformation of the VQS to a landed fish and the exchange of VQS on a market. The complicated paperwork, regulations, and rules governing the three stages and the connections between them create the possibility for people to provide a service to rights holders with expert knowledge[2]. For example, in-depth knowledge of the exchange-value and its regulations—especially when leasing is involved—forms

---

[2] "The use-values of commodities provide the material for a special branch of knowledge, namely, the commercial knowledge of commodities." (Marx 1976, p. 126)

the basis for experts, such as accountants or quota pool managers, to manage quota and advise people on VQS transactions. On the other hand, knowledge about use-value forms the basis upon which talented operators of fishing activities can live.

Ethnographies of Exchange

In the following ethnographic section, the focus is on the exchange of VQS. I have distilled material from interviews and field notes that deal with the exchange-value. I have already pointed at some economic strategies and aspects inherent in the VQS system. Fluctuations in exchange-value were the basis of strategies related to buying and selling of VQS. In itself quota speculation is a distinct strategy, but more importantly it has become one of the considerations VQS holders have to take into account, at least at some point in their life-course. What will the value of quota be next year, or in 3 years? Should I buy now, or lease, or sell? The specific combination of leasing prices and the financial market made it possible in principle to invest in VQS in order to lease to others. What some pay to lease "fish" can be higher than the long-term costs of actually buying the quota. But in turn, leasing prices are also dynamic and seem to change in relation to supply and demand, wind, weather, and fish behavior, as well as annually because the allocations have to be used within the year. Beside the right holders, there is a set of experts with knowledge of the transactions and their regulation that can provide help and advice. Thus, the qualities of the VQS as a commodity involve new thoughts, professions, and behaviors for the people engaged in commercial fishing. Included in these new situations and behaviors, and where exchange-value is particularly important, is the buying and selling of VQS, which marked the early days of the VQS system when it was first introduced in 2007.

About Selling

In 2011 I interviewed Ole[3], a retired fisher in central Denmark who sold his vessel and VQS in the first month of the VQS system, January 2007. Ole began fishing when he was 18, back when "the only thing you had to do was to catch some fish" (Personal conversation, November 2011). He bought his vessel in 1970. It was a 14 m long wooden trawler that had been built in 1963. In December 2006, Ole received notification of his allocation of VQS. For that reason he called it "a little Christmas gift" (Personal conversation, November 2011), and a few weeks later Ole sold his vessel and VQS to two local fishers who "split" his VQS between their two vessels. When I ask him why he sold, Ole lists a range of reasons for selling:

---

[3] All the names in this chapter have been altered.

- As the allocation was based on the principle of catch history, Ole's allocation reflected his fishing activities in 2003–2005. In that period he had been fishing away from home for six to eight months each year. He had been fishing on the west coast of Denmark in the spring and around Bornholm in the most eastern part of Denmark in the winter. Although he had been trying to focus his fishing in Kattegat and the Western Baltic, the two most adjacent catch areas, the declining quotas there forced him to participate in the more lucrative flatfish and cod fisheries west of Jutland and in the Eastern Baltic. The low quotas had forced him to more distant areas, and just a few weeks before he received news on his VQS allocation, the 2007 cod quota was cut down by another 15% (Personal Communication, November 2011).
- In addition, Ole's herring partner—the owner of the other vessel he was pair trawling with for herring—had already sold his vessel, so Ole would have to find a new partner and vessel to arrange the herring trawling with. When he sold his vessel he had health problems, attributed largely to a long life with hard physical work onboard a fishing vessel. In this part of the interview his wife entered the kitchen where we sat and commented on the vessel owners who are selling: "They are worn out. A great part of them are." (Personal conversation, November 2011)
- For Ole, the individual quota shares reduced flexibility to switch between fisheries, for example targeting plaice if the sole or cod fishery failed. To do that he would need to invest or lease in further VQS. As explained above, the principle of historical catch resulted in allocations in distant areas, which he would have to swap or trade in some way to get VQS closer to his local area.
- Ole stressed that the VQS only represents a share, and you never know what the annual allocations will be. In addition, you will never know the landed value (auction prices) of that annual allocation: "There are two unknowns when you buy. How many tons you actually get and what it is worth." (Personal conversation, November 2011).
- Ole was 59 and would only have a few years to pay back any investments: "Most people would probably keep away from getting indebted when they can see the end date." (Personal conversation, November 2011). This point should be understood in relation to the insecurity in annual allocations and landed value mentioned above.

A range of different reasons explain why he sold his vessel and VQS. For Ole, it was a chance to leave a practice that had him fishing away from home more than 6 months each year. He was, in other words, not an inefficient fisher. Perhaps Ole could have sold his initial allocation and instead acquired quota for the nearer catch areas. However, such a "swap" would have left him with VQS in an increasingly fluctuating fishery with lower value and more hard work. His health was marked by the strenuous work on board for many years, and his age meant that making large investments would be attached with greater risks. However, when I met with Ole in 2011, he was working at the harbor taking care of the incoming landings of fish. Another factor was that Ole's herring partner had already sold, so Ole needed to find a new partner in order to continue pair trawling for herring. As many fishers' colleagues and former partners sold their VQS, the social fabric of the sector changed, and those left were forced to find new collaborators and partners and even new harbors to land in. As we know, because of overcapacity, TAC was split up among the many VQS holders, leaving many fishers with too little quota to make a living, which forced the initial VQS holders to either invest or sell. At the same time, many fishers were reluctant to take loans because of age, economic thrift or ideology. They were not going to pay for something that used to be free. All this left Ole and many others with a hard choice: "If you wanted to make a living out of it, you would have to go out and buy" (Personal conversation, November 2011). As we have learned, Ole decided to sell his vessel. From meetings with other vessel owners, I have discovered similar reasoning:

- In central Denmark, another vessel owner who sold his vessel mentions other reasons. Most of all, he was disillusioned with the numerous regulations and restrictions put on fishing activities. The "days at sea" regulation was a heavy burden, and with the announcement of a closed section in his fishing area he decided to quit and try to find a job on shore. Another reason he gave was his age. He had just turned 55, which allowed him to put the income from the sale into a pension fund, and thus he could reduce the tax burden on his income from the VQS sale. When I met him again in 2011 he was fixing a boat and planning to start up a new fishing operation. Finding another job on land turned out to be difficult (Personal conversation, December 2011).
- In contrast to the two above who were running medium-sized operations, in the Eastern Baltic I met a skipper and owner of a large trawler with substantial catches who also invested in buying VQS in 2006 and 2007 (Personal conversation, December 2011). In this case, he was selling his considerable shares of VQS to an even larger company, because he believes that "the time is right now" (Personal conversation, December 2011). He will continue fishing on the vessel he sold, but now as a hired skipper for another company. He tells me that the price was right, even though he will go on fishing for a few years. With the money in the bank (or pension) he would be relieved from the uncertainty of the changing quotas (TACs) and would have "secured his part," the buyer later explained (Personal conversation, December 2011). With quota selling for one-third of the price a year later, he most probably made a good decision at that time.

Michael's case is an example of a strategic sale where the current and future value of the VQS was taken into consideration when selling. In contrast, I have also met VQS holders who sold because they found it too hard to make a living out of their VQS and left the fishing industry to find jobs and work elsewhere. What we see from these qualitative insights are a wide range of reasons to sell VQS, all rooted in a specific situation or practice but at the same time relating to the exchange-value. It is hard to argue that these sales are irrational, but they cannot be reduced to a single rational relation between more or less efficient fishers. Rather, a whole range of factors are influential at the individual level, and financial access and risk aversion seems to be crucial.

During the first few years of the VQS system the fleet was reduced by 25 % (Schou 2010). This means that out of a total number of almost 1000 vessels initially allocated VQS in 2007, more than 200 vessels with their respective VQS were sold by their owners. Seen in regard to the many operators and the high average age (50 years in 2007) this is not a surprising finding. Ole was one of the owners who sold, and it is worth noting the relation between the large number of retiring vessel owners and the large number of entrants in the 1960s and 1970s, as many from this age group reached the retirement age around 2007. Health and age is a driving factor for sale, while some sell because they have grown tired of the industry. Further, difficulties with changing regulations and increasing administrative burdens push people who feel "run down" to sell. This expression covers a range of factors from economic performance to everyday wellbeing as well as weakening social networks.

As the map (Fig. 3.2) in the previous chapter indicated, a number of communities experienced half or more of their commercial vessels being sold (in some places in a matter of weeks or months). The rapid decline and change in the social fabric is also a factor in these decisions. The presence of fewer individuals to run and finance the landing facilities is a push factor for selling itself. Behind all these factors—age, health, declining networks, and frustration over an ever-growing set of regulations—are the fluctuations in VQS and the insecurity that arises from this. As Ole said, you never know how much you will get and what you will get for it. Some of the reasons for selling are personal and some are based on more general considerations, but most of them are embedded in specific situations and fishing practices besides being linked to the VQS as an economic asset.

On the other hand, the rapidly rising amounts of money paid to VQS holders could be described as a pull factor. The combination of a pressure for VQS—a race for quota so to speak—and accommodating banks triggered a race that pushed prices up. This was a rapid process, where individual gain often had to be balanced with a loyalty to the local community as the decision to sell affected not only the vessel owner but also crew and their families. The fact that Ole needed a new partner for herring pair trawling perhaps pushed him to sell when a tangible offer was made. Had he waited, he might have seen the flexibility of the new leasing system and perhaps would have found a solution to his situation, for example, by leasing out his herring VQS. On the other hand, if Ole and people in a similar situation had not sold, would the leasing market have been as dynamic? This is of course bare speculation, but what is certain is that the fleet reduction experienced between 2007 and the current day encompasses an ocean of individual stories, all of them worth more attention than I can give here. Each individual sale is a story rooted in a fishing practice.

The Sale

Returning to Ole and his story, he went on to explain that the actual sale was initiated by people from outside his community who asked if he was selling or not, while he in fact wanted the "fish" to stay local.

I got questions from the outside. When one of them was made tangible as 1.8 million, I asked locally if anyone was interested in buying at that price, to keep the fish local. (Personal conversation, November 2011)

Ole then sold his vessel and VQS to two local fishers who "split" his VQS between their vessels, and so the "fish" stayed in the community—at least for a while. This way Ole balanced his individual gain with his wish to support his local community. As we know, Ole sold only a few weeks into the new system, so I ask him if he should not have waited some time to see where the prices were developing. Here is what he answered:

> Sure, and had I waited two years I might have received less. There is no need to worry about that. If we were satisfied with what we got then that is the situation. It might very well be that it goes a bit up and then comes down again, then they cut the cod quotas and then increase the sprat. We had such a mixed allocation. I am not worried about it […] When you have signed it is done. (Personal conversation, November 2011)

This was his reasoning around the exchange-value. Ole had another fisher as crew on his boat, and I ask what happened to him, and whether it was fair that Ole received the income from the sale while the crewman just lost his job. According to Ole, the investments and the financial risk he assumed justify the distribution of VQS to him only. The crewman could leave from week to week without obligations, while Ole was financially bound up with investments in gear and vessel. Luckily for the crewman, he quickly found another job as a construction worker; but this is not the case for all. As many justifications can be found for the sale of VQS, the same is the case for buyers. Investments in VQS are rooted in a practice that explains the acquisition, but surely considerations in regard to the fluctuations must also be present.

To Buy or Not to Buy?

In contrast to Ole, who was approaching retirement, I have interviewed a young man who hoped to be an independent skipper on board his own boat. As a young and new independent skipper you receive a small and short loan of VQS from the state (an arrangement which will be explained in the next chapter). For a new skipper, there are two ways to get access to the commercial fishery: you can either lease or buy VQS. But in this case, buying VQS is currently not on the agenda. He explains why when I ask him if he is considering buying quota:

> Fish? I am not thinking about it. It is a lot of money to invest in something that quickly drops and falls, it is like buying financial stocks. They rise and fall too, you cannot count on that. (Personal conversation, January 2012)

Clearly the fluctuations in prices are enough to keep him away from buying, but he also sees VQS as a dangerous investment that will take a long time to pay off. Here he is talking about those who buy:

> If they buy a kilo of cod at 130 kroner, it will take many years before you have paid that off. Then people say we can always sell them again. Yes, but if you are not about to sell them, how many years will it take to pay them off? It will take at least 12 years before the cod is paid off, because you only have them once a year. (Personal conversation, January 2012)

In the quote above he is talking about the two perspectives on VQS as a commodity. First, as an investment that will keep its value and be an economic asset later in a sale; and second as a means of production you only have "once a year." In the first, VQS appears as an exchange-value, while in the latter perspective it is a use-value that should provide income as a necessary part of production. The new entrant in this case is clearly focused on the use-value aspect, perhaps because of his age, and I ask him if buying VQS could be compared to buying real estate as an investment, because it will keep its value. However, in his opinion, venturing into VQS acquisition is more like entering the financial stock market than the real estate market: "It is the same with fish [VQS], it is exactly the same, there is no difference" (Personal conversation, January 2012). Seen as a use-value, he can make an income from leasing VQS, while investing in VQS requires him to take out a loan and pay the bank instead, something he is reluctant to do. In the next chapter, I will go deeper into this mode of operation and examine the way he structures his operation.

Turning now to those who invested rather extensively, my interviews indicate that *leasing in* makes little sense as a strategy for them, especially in times when the interest rate is low:

> [*The low interest rate*] is also why we bought Michael's [*VQS*]. It was because we had to go on the leasing market to get enough fish. Then we went over the numbers and we could see we had a much larger rate of return by owning them. (Personal conversation)

It should be noted that low interest rates are not a generalizable experience. Many fishers experienced high interest rates and rising financial costs since 2008 as a hindrance. But for some, access to financial capital is cheaper than for others; and therefore there is also a place for an economic strategy based on taking out a loan to buy quota, then leasing out the "fish." The interest rate span can vary as much as from 2 to 12%, even though it is the same commodity they are buying (at least these are the two extremities I have discovered in my fieldwork)[4]. These conditions make a significant difference to the financial costs (see the example in the footnote). In addition, the VQS system created collateral that changed the financial situation radically for the vessel owners. Here explained by a skipper who runs a one-man operation:

> Back then we had to ask and beg to get a loan from the banks in the region. But then suddenly, when they found out how much money we had, we had no problems and we could buy as much as we wanted. (Personal conversation, December 2011)

The privatized exchange-value both created the possibility to commence investments as well as a security in the new investments. This enabled even small opera-

---

[4] Comparing two 7 year loans of 1,000,000 DKR at 2% and 12% interest result in annual payments of 153,000 DKR and 210,000 DKR, respectively. In total, the financial cost of the 12% loan is in total more than 400,000 DKR higher.

tors to expand if they were willing to take on the financial obligations. Of course, it also meant that in the case of a sale they would be able to cash in the new value.

> All of a sudden by the end of the year when the value of the boat was estimated—it had increased by two or three million in value, the quota, in one year. Mikael, the accountant, said that was something of a change. The banker came and shook my hand and congratulated me, 'congratulations with the title', I was one of those millionaires. (Personal conversation, December 2011)

The introduction of individual transferable fishing rights enabled both selling and buying for smaller operators. However, larger company structures coupled with the security of fishing rights provided some companies with an even better opportunity to invest and expand. This is still the case today—even though the banks are more reluctant to give loans now than they were prior to the credit crisis in 2008. A critical point here is that the banks often had several vessel owners as customers, and through that banks could technically influence the development of the fishery by giving and denying loans. As will be discussed later, access to credit might be a better explanatory factor of market dynamics than "the most efficient fisher."

What can be shown is that large companies were the first to "buy up" other vessels, a process that started even before the Danish government decided to introduce the system:

> Then there were talks about this system. Then I read a lot of reports from Iceland, Scotland and Germany, which had introduced it before us, Individual Transferable Quotas, there were a lot of reports on it. I could see, and my partner could see, the possibilities in it. Either we would have to sell or we should go on and buy. Then we started to buy before it was decided, it was decided in October 2005, and at that time we had already bought the first two vessels. (Personal conversation, December 2011)

As we see from the quote, the partners researched the system in advance and gained a head start in the process. They reacted quickly to the situation and benefitted from their research, not only compared to people like Ole but also to others who bought later when the system was introduced. This underlines the knowledge, research, and information aspects of dealing with the exchange of VQS. Without an analysis of the situation and future development, buying VQS is pure speculation. Therefore, the company did research on the development of market-based policies in other countries and knew that quota prices would probably rise and concentrate in the hands of fewer operators. It would also allow the company to secure more stable seasons and work conditions. Based on their own research, they decided what to do. But the information could also come from other sources, for example the local branch of the fishermen's organization.

A negative example of this is the advice that Bornholm vessel owners were given by their local branch of the fishermen's organization when the system was being unveiled. In contrast to the company mentioned above, which bought very early, the vessel owners who were members of the Bornholm branch were advised to wait until all the rules were in place, around the beginning of 2007. This resulted in significant stress and the loss of opportunities to buy early, while others from far-afield started to buy the vessels around them. Here a vessel owner explains the development:

> [...] we were just standing and watching, what is happening, and our chairman is just watching as well, and they were just around us buying, and we were offered one price after another. And we were only talking about 1.2, it was crazy, today they can get 3 million. What they sold for 1.2 back then, it was giant money, a boat rose from 450,000 to 1,200,000 and people got tempted and sold. (Personal conversation, December 2011)

The rising bids, their high average age, and the insecurity as to future developments meant that when exposed to the high exchange-value, fishers were tempted and sold. Afterward, some fishers bought back their vessel, now stripped of VQS, and continued fishing, using the leasing system to gain access. This seems at first to be a contradictory strategy, but the reasoning behind it can probably be found in the independence they gain from the financial markets and from exchange-value fluctuations. This point will be elaborated in the next chapter. At the other end of the scale, some have invested and now lease out to those without any or enough VQS. The skipper interviewed above from the large company—now speaking of another large company—explained the logic and paradoxes of that strategy:

> That is because in recent years you have been able to lease out fish and make a living from that. But that will change now. It looks like the amounts are declining and the fishers are becoming fewer. You cannot just take away all the fishers and then still think you can lease to them at the same price. He is buying all the vessels. I do not understand his strategy. He wants to lease out and make money from that, because the interest rate is so low. (Personal conversation, December 2011)

Here we see another factor worth highlighting. Lease prices are seemingly highly sensitive to supply and demand. Since the amounts of "use-values" (the TAC) are declining, the vessel owner will not have enough use-values to both run his own operation and lease out at the same time. On the other side fewer are active in the fishery, so there is less demand for the (excess) VQS. The vessel owner referred to in the quote had at the time bought around 40 vessels; so it would be fair to say he had contributed to fewer operators, who were at the same time also his customer base.

Leasing or Losing?

The possibility to lease VQS was introduced to help fishers avoid discarding, by allowing them to lease small amounts of VQS when catch compositions were not as expected and to promote cooperation between operators. But for many the leasing system is the only way to get access, and large amounts of VQS are leased every day in the so-called fish pools. As described above, leasing is the transfer of VQS for the remaining part of the year. Where buying and selling can be described in almost neutral terms—or at least as actions between willing buyers and willing sellers—there seems to be more inequality involved in leasing. In the case of leasing, a group of vessel owners and fishers without VQS or with too little VQS and are consequently in need of something others have. On an Internet forum for fishers, leasing prices and the leasing system were recently debated. The debate was started by a critical new entrant in the fishery:

> I do not think there is any help for a new entrant today with the VQS system. I cannot make money by giving such a high rent for fish!!! I think it is grotesque that several of those that have sold their vessel, still hold the quota, even though they are working on land. Several of the big ones just buy more quotas to avoid the tax, while we small ones are losing out. With a bank interest rate of 12–13 % it is impossible to make a future living as a fisher with your own boat [5]

What initiated the debate was high leasing prices, in this case, 2 kroner in lease for plaice, a fish that on a good day yields 8 kroner in landed value. In itself this is not a critique of the system but of the relation between the leasing price and the landed value. This situation is of course most delicate for new entrants and others who are more dependent on leasing. The debate was joined by another "new entrant" who suggests that the first poster chose the wrong strategy and should have engaged in a "generational handover" instead. In such a handover the fisher enters into partnership with an established skipper and vessel owner in order to take over that company bit by bit, or to run the operation in exchange for 10 % of the company. This suggestion reveals another aspect of the exchange-value problematic, the exit, or the handover. In many ways, this is a conflict similar to the dilemmas in the previous chapter, where the market creates tensions between community values (handing over to colleagues, family, neighbors, etc.) and the individual gain that can be achieved from the sale. In addition, the handover or exit is a financial problem, as with larger and larger accumulations of VQS there are fewer potential buyers with the required financial power.

The Internet forum, which is not known for its gentle language, is then joined by a third person who reacts to the previous suggestion to enter a generational handover:

> Just a little question. You say you got good help, but can you look in the mirror and say 'I am skipper and I am in charge'? Probably not. There is likely some ship owner who sees you as a good way out of fishing when he is fed up fighting his giant debt. [6]

What we see here is the clash between two strategies or visions of how to be(come) a skipper and involved in the fishing sector. One is to enter into partnership and joint ventures, while the other is to seek independence from others in respect to ownership and managing the operation. Again, we will look closer into the differences in the next chapters. There is also in the comment above a critique of the social class of owners who exploit the fishers without VQS. Finally, another fisher enters the debate with his own critique and suggestions for fixes:

> No matter if it is a small or large vessel, the leasing price is way too high at the moment. I myself need a little extra quota to make ends meet. But no matter if it is coalfish, cod or plaice it is totally crazy with those prices you have to pay to lease fish, especially when the fish [landed value] is not already too high on the auction. The price should be set in relation to the auction price, and perhaps it would be a good idea to let the pools organize the prices. Because with the high price we have now, there is more discarding, and when the year is

---

[5] http://fiskerforum.dk/debatforum/debatforum.asp?mode = viewmsg&Id = 3854&ForumID = 10&Showsub = 3859 (Accessed September 15, 2012)

[6] http://fiskerforum.dk/debatforum/debatforum.asp?mode = viewmsg&Id = 3854&ForumID = 10&Showsub = 3859 (Accessed September 15, 2012)

Fig. 4.3 Talks at the harbor. Leasing does not only take place via the internet based market, but also more informal deals are made, often to help each other locally with specific needs or excess quota shares (Photo: Jeppe Høst)

> about to be over, large amounts will be put for lease at low prices, and what the hell will that help when it is the 24th of December?[7]

In many ways the quote above sums up the numerous aspects of leasing. What is at stake in this discussion is the leasing price of the VQS, and this is compared to the landed value, the value of the fish commodity at the auction. The relation between the two decides if you are "working for free" or if you are making money. The fisher suggests that an organizational structure should set the prices, which should be controlled and set in relation to the auction prices; and thus individual gain and market fluctuations would be under some kind of community control. The quote also indicates that the high leasing prices can result in "over flooded" markets in December, when too many "fish" are put on the market. As we will learn in the next chapter, leasing also takes place informally between the operators as part of their daily interactions (Fig. 4.3), sometimes at reduced prices, while the internet-based market sets the overall market prices.

All in all, the behavior around leasing, buying and selling revolves around four different value aspects of the fishing practice. These four different values are of course the use-value and exchange-value of the VQS, but also the landed value of the fish commodity and the leasing price (or rent value). When Ole, the vessel owner who sold, said that "you never know what you will get and what it is worth," he was talking about the total use-values and the value of the fish at the auction. Likewise, when the fisher discussed the leasing system in the Internet forum, the correla-

---

[7] http://fiskerforum.dk/debatforum/debatforum.asp?mode = viewmsg&Id = 3854&ForumID = 10&Showsub = 3859 (Accessed September 15, 2012)

tion between the leasing price and the landed value were the most critical aspects. Where the use-value aspect—the specific use of the VQS—transforms the VQS to landed value, the exchange-value is revealed by leasing and permanent transfers. These are linked of course, as we saw a VQS that is not *used* before the end of the year is of little exchange-value. However, one question is left unanswered: what determines the value of the VQS? And what is the value that is actually exchanged in these diverse transfers?

Value and Value Fluctuations

It is not an easy task to determine the value of the VQS. As already mentioned, a whole range of factors influence the fluctuations in VQS prices: landed value, oil prices, catch rates, financial costs, supply and demand, and so on. Shipbrokers, quota pool managers, and fishers know and talk about the prices, but is there a logic defining the value? The following is an excerpt from an interview where a skipper and I discuss the value and prices of the VQS:

> Author: What sets the prices on the quotas?
> Vessel owner: The landed value of the small pelagic is insanely high at the moment. That means the market value also is higher. Those follow each other.
> Author: That is a factor?
> Vessel owner: And then supply and demand.
> Author: And what about how much you can borrow in the bank?
> Vessel owner: Yes, that is another side of it.
> Author: Oil prices?
> Vessel owner: Of course that is a factor. […] We are forced to use diesel, but the diesel prices are historically high. I have never experienced such high oil prices. It is at a level now where the cost for oil makes up a quarter of the fishing [*this probably means that oil costs consume one quarter of the landed value*].
> Author: You cannot really know if the TAC for that species you invested in will go up or down?
> Vessel owner: I just mentioned sprat to you. It is going down. That is why they are busy trying to acquire some more rights so they can fill up their vessels and get an outcome from making some trips.
> Author: Will sprat [*VQS*] then be more or less valuable?
> Vessel owner: It will be worth less, because the TAC will be reduced next year.
> Author: But you say that the demand will increase?
> Vessel owner: Yes, then the demand will increase, because the large vessels can land a certain quantity, and then they can see their amount is declining. But for it to pay off for them to sail all the way up in the Eastern Baltic, all the way to the other side of Gotland, which is two days of steaming, then they need to be able to fill the vessel. Because of that it makes sense for them to buy x amount of fish, to be able to make the trip and return with a full ship.
> Author: But the value of the right will not increase?
> Vessel owner: It will, but that is according to supply and demand. (Personal conversation, December 2011)

As the communication above indicates, it is not easy to identify the value of the VQS and understand the dynamics behind it. Its value will go up and down at the same time. Perhaps a distinction between value and price could help to explain a lower value but higher price due to demand. Returning to Marx we might find inspiration on how we can conceptualize the value. Marx's most widely known theory of value is the labor theory of value. In this he argues that the value of commodities is the *socially necessary labor time* required to produce that commodity. But if the VQS is not a commodity produced by labor, then what is its value made of? Since the value of the VQS appears to make up on average 80% of a fishing operation, this is also a rather important question. I have argued that the VQS is a politically established commodity marked by its properties as a political management tool. Could we then argue that it is the administrative labor that makes up its value? In other words, it would be the combined administrative and bureaucratic work put into maintaining the quota system.

Besides the labor theory of value, Marx also had another set of value theories concerned with resource rents that he applies to land and other natural sources of income. The concept of rent and of resource rent is well-known in fisheries economics (Grafton 1995; Squires et al. 1995). In fact, in relation to capturing rent, market-based fisheries management is seen as one of the most efficient management methods:

> The theory behind ITQs suggests that in certain fisheries they can be an effective management tool to prevent rent dissipation and can increase the returns to fishers and the resource owners. (Grafton 1996, p. 17)

The argument made is that with free and equal access, rent is not captured. However, with the resource as private and individual ownership, it can be operated efficiently and a rent can be captured. This is for the benefit of the fisher as well as for the states that can choose to capture the rent. However, the social origin of this rent is often not problematized or examined. In Marx's rent theories, rent is explained as something coming out of a social relation between the land-owner and the land-user. In other words this is a monopoly rent—a price paid for the use of a resource to the owner. Will this rent value theory be able to explain the complex fluctuations and the relation to the landed value and production costs? Below I will—in a somewhat compressed manner—discuss the two approaches to see how they contribute to an understanding of the exchange and trade with fishing rights and the underlying value that is traded.

Labor or Rent

So far, I have analyzed the use-value and exchange-value as two aspects of commodities. There appears to be a correlation between the use-value and the exchange-value, since the use-value is the right to catch and land that share of fish and thereby create fish commodities for a market. This relation also came up in my ethnographic material, where decisions around selling, buying, and leasing were both rooted in a concrete fishing practice and based on considerations around the fluctuating exchange-value. But since the exchange-value can be expressed in money, the two are also independent of each other. It is as a part of the exchange situation that the VQS can be expressed in other commodities, most often in money:

> We have seen that when commodities are in the relation of exchange, their exchange-value manifests itself as something totally independent of their use-value. (Marx 1976, p. 128)

The heterogeneous use-values are somehow exchangeable and can be measured against each other, despite their different qualities and the different types of work put into their creation. But in order for two commodities to be exchangeable, Marx further argues, they must have a value and their value must be defined by something outside them both:

> Both are therefore equal to a third thing, which in itself is neither the one nor the other. Each of them, so far as it is exchange-value, must therefore be reducible to this third thing. (Marx 1976, p. 126)

This third thing is not money but, Marx argues, the fact that when stripped from everything else, all commodities are products of human labor. Labor is what makes up their value, Marx argues. But in order to understand their exchangeability, we have to look at the difference between the *concrete human labor* that produced the use-value and *abstract human labor* needed in a society to produce the commodity. When exchanged commodities:

> [...] can no longer be distinguished, but are all together reduced to the same kind of labour, human labour in the abstract. (Marx 1976, p. 128)

Seeing the twofold aspects of labor as both concrete and abstract is the critical analytical move that allows Marx to find the value of commodities. There is individual labor and a general comparative labor. Concrete labor is what actually produced the specific commodity; but abstract labor is the amount of average skilled labor it would take in society to produce that commodity, measured quantitatively. The concrete labor could be expensive and slow, but the value of the commodity will be measured in abstract labor, the average labor power needed to produce that commodity in a certain society. It is this abstract average labor that we use when we compare the value of two commodities.

Therefore Marx finds the value of a commodity in what he defines as the *socially necessary labor time*, which is human labor time in the abstract.

> The value of a commodity is related to the value of any other commodity as the labour-time necessary for the production of one is related to the labour-time necessary for the production of the other. (Marx 1976, p. 130)

The above theory of labor value is an essential part of Marx's analysis of the capitalist mode of production. It is through the labor process that the surplus value is created and not as an instrumental yield on capital and technology. However, as mentioned above, fishing rights are not created by human labor, and by using this value theory we can only explain the landed value of fish. Can this indirectly explain the value of VQS? Is the value of the VQS related to the landed value, as it is the human use of the VQS that gives it value? This is tempting but does not fully explain the value of the VQS. How many years of human labor does a permanent transfer represent? Why is it four in the Baltic Sea and 12 in the North Sea? Exactly which fraction of the landed value constitutes the leasing price? Here we could point to loan conditions, tax exemptions, and fishing conditions, but these would contradict the core of the argument, that labor creates the value; and it would not sufficiently explain the regional differences. These would suggest instead that the value is a network of political, administrative, financial, and concrete relations of production. In many ways the labor theory of value does not provide a full explanation of the value in regard to fishing quotas. If we continue within the Marxist arguments and theory, the theory of rent addresses the above problem in a different manner.

Monopoly Rent

Marx does not use the term *monopoly rent*, but rather the terms *ground rent, absolute rent*, and *differential rent*. In this compressed discussion I will use the term *monopoly rent* to stress that social relations and ownership are the basic source of this rent (see also Ramirez 2009). Marx discusses the rent question in relation to a complex debate on land rents in agriculture, but he also uses the example of a waterfall to explain the principle (Marx 1981, "Capital Vol. 3"; see also Ramirez 2009). I will begin by outlining the latter example, which essentially forms a good starting point for understanding the idea behind monopoly rent, then turn to the monopoly rent in fisheries. Imagine two factories producing the same commodity. One uses steam power and the other is powered by water stream. The price for the product they sell is set by the market (including an average rate of profit), and the market price is thus the same for the two producers. However, through the use of water stream one of the producers can produce a *surplus profit*:

> [...] because their commodities are produced, or their capital functions, under exceptionally favourable conditions, conditions that stand above the average level prevailing in this sphere. (Marx 1981, p. 780)

This surplus profit is like any other profit: it is derived from the use of human labor and is the difference between the individual price of production and the general price of production. The surplus profit comes from the fact that the labor to produce the steam power (and coal) is not a cost for the producer powered by the waterfall. Both of the producers can use innovations and expand production, but the two costs of production will, in principle, not equal out with time. The water-powered producer still has an advantage—at least until something makes the waterfall useless in the production. This advantage does not come from capital nor from labor, but from a connection of these and the exclusive use of a natural force. In other words, it is a resource that is limited and that "is available only to those who have at their disposal particular pieces of the earth's surface" (Marx 1981, p. 784).

> Those manufacturers who possess waterfalls exclude those who do not possess them from employing this natural force, because land is limited, and still more so land endowed with water-power. (Marx 1981, p. 784)

The landowners can apply capital, but they can also prevent the application and use of capital on the waterfall. They can lease it to others and charge a ground rent. It is this exclusive situation I refer to as the monopoly situation. Furthermore, because of this, *profit* and *surplus profit* can be conceptually separated:

> The surplus profit that arises from this use of the waterfall thus arises not from the capital but rather from the use by capital of a monopolizable and monopolized natural force. Under these conditions, the surplus profit is transformed into ground rent, i.e. it accrues to the owner of the waterfall. (Marx 1981, p. 785)

In this step, Marx isolates surplus profit from the capital and profit in order to create the notion of ground rent. Ground rent can be separated from the profit because the surplus profit does not come from use of capital alone. This makes possible a situation where the owner and the user of the waterfall are not the same, a situation where a landowner extracts the ground rent. However, should the capitalist also be the owner of the waterfall, he will pocket both a surplus profit as a landowner and the average profit as a capitalist.

The ground rent is in Marx's terms a differential rent, because it always arises from the *difference* between the individual production price and the general production price. In short, ground rent applies to a situation with owners of a monopolizable natural force that can benefit a producer through capital and labor, which results in a surplus profit that can, in turn, be transformed into a monopoly rent. In agriculture there is not only a differential rent, but also an absolute rent. In Marx's understanding it was the least productive plots of land that set the market price. The difference between the market value and the production price on these meager plots of land is the absolute rent, whereas the surplus profit derived from better plots of land is a differential rent (see above). Still, it is only in the moment that human labor is employed that value is created, which then can be extracted by the landowner as rent (based on the monopoly situation).

Monopoly Fisheries?

In the case of commercial fisheries managed through transferable private property rights, we see a similar situation with an exclusive ownership of a natural force and—like agriculture—a sector with fluctuating market prices. The rent theory should then state that the lease price is the extraction of the surplus profit by an owner from a user (leaser). This process is dynamic, and the leasing market is where this is negotiated. Here, prices and amounts go up and down as a constant struggle between leasers and owners in relation to the fluctuating market prices. The theory also proposes that the owner of the VQS can separate his income from the VQS in ground rent and profit: rent originating in the exclusive ownership of fishing rights and profit made in the application of capital and human labor. In the leasing situation, this is clearly visible as the two beneficiaries are separated and the monopoly rent appropriated by the owner, while the income from the use is appropriated by the producer(s). But in theory, a skipper engaged in fishing her or his own VQS will have both an income from production and from a resource rent. This is harder to see but might explain why the boat share—the share of the income that goes to the vessel owner—tends to go up in market-based fisheries management. The monopoly rent is integrated into the boat share. Some companies lease to themselves, in which case the rent can be subtracted like oil costs in advance and therefore not paid as part of the boat share[8]. This technically separates the rent from the income.

The rent is generated by the difference between the price of production and the market price for the commodity. In this regard it is difficult to define exactly what determines the market price. The seafood market is, despite the perishable nature of its commodity, a global market, with both international supply and local landings on one side and international demand and distribution on the other. What we can suggest is that when the market prices on landed fish go up and more vessels can go to sea with a good economic outcome, the leasing prices will also go up. Because according to the market pattern the "owners" will seek—or can seek—to extract the surplus profit by setting the leasing prices higher. This explains the connection between the leasing cost and the landed value, which *appears* to be controlling the lease prices.

The rent theory suggests that this is not so, but instead that it is the social relation and the *monopolists*' work to appropriate the surplus that sets the prices. The value in this understanding is the monopolized resource and the rent this social situation yields. Following Marxist inspired logic, this seems to be providing an explanation of the value when analyzing the lease of fishing rights. It is the monopoly situation, with private property rights, that creates rent, while there would be no value without human labor. In that respect, we need both the rent theory and labor theory to explain the value of VQS. Resources not under human use would yield no value and consequently no rent, which explains the flooding of leasing markets in December, when fish owners seek to secure some rent before the expiration of the annual VQS. The power balance in the leasing situation shifts slightly by the end of the year in advance of the leasing part.

Next we must ask if the rent theory can also account for the value of the permanent transfers. If the parallel to land above can be drawn out further in order to explain the value of the VQS in permanent transfers, then we will have to look at the VQS as capital that yields a rent. As capital, it can be circulated and exchanged and therefore must have a value that can be compared to other values. In other words, plots of land can be sold, and something determines the value of that land. Its value is therefore its equivalent in other forms of capital or money. The rent theory offers a way to calculate the value of that capital, which is dialectically determined by the leasing price and the average rate of profit. We can look at a numerical example from the Eastern Baltic cod in 2001, where the average lease price was 3 DKR per kilo. If the average rate of profit is set at 8%, then the numerical value of the VQS

---

[8] A few months into the new system there was an agreement between the owner and labor organizations that the lease costs could be subtracted in advance, in the same way as oil and other costs of production. In this way it is not paid as a part of the boat share like the vessel and gear.

would be 37.5 DKR[9]. That would be the amount of capital that would yield a profit of 3 kroner if it was applied in another sector. Viewed in this way, the value of the VQS is a dialectic relation between rent and average rate of profit—it has in other words been capitalized or financialized. Functioning as capital it can be circulated and exchanged, while its annual yield and permanent value is still dependent on the actual use of the resource. It can be compared and integrated with other productive sectors, and therefore, it can be conceptualized as capital and operated for purely commercial motives. Thus, the rent theory offers a coherent explanation that integrates labor theory in order to explain the value of both leased VQS and permanent transfers of VQS. In contrast to the common rent theories in resource economics, rent is here explained as a social relation and not as something just coming from an efficient use of the resource.

Using rent theory we could evaluate the VQS numerically. The example above is from the Eastern Baltic and is not far off the prices paid on permanent transfers (42 DKR). More importantly, what is offered is a theoretical connection and principles that can explain the value lying behind exchange situations in market-based fisheries management. Exploring Marx's theory of rent was an inquiry into how capitalism would operate in agriculture in combination with a social class of landowners. While the rent theory in combination with the labor theory of value is useful in explaining values in the VQS market, there might be other noncapitalist reasons to lease or invest in VQS and other ways to be a producer. The actual use of the VQS and the different modes of operation will be the objects of discussion and analysis in the following chapters.

Conclusions

This chapter has shown that the introduction of market-based fisheries management brought a new exchange-value into the vessel owner's practices. This new value propelled a dynamic trade, which in turn led to a redistribution of the fishing rights described in the previous chapter. The exchange-value aspect could be separated into a distinct task, where quota pool managers, ship-brokers, and accountants became experts of exchange. However, for most vessel owners the change was more about a new dimension in their practices, where prices on VQS have to be discussed and the leasing market monitored. Ultimately, the fluctuations in exchange-value are an important part in fisher's long-term strategies. Decisions on when to invest and when to sell and retire were directed by market developments. The ethnographic research pointed to different financial conditions as having a great impact on the reasoning behind buying and selling.

The economic literature proposed that the most efficient fisher would buy fishing rights from the least efficient fisher. In contrast to this, the above analysis proposes

---

[9] To find the value of VQS as capital we can use the equation: leasing price/average rate of profit = value of VQS.

that if one principle is to be used to explain the developments of the fishing fleet under VQS, it should probably be competition in financial markets rather than the triumph of the most efficient fisher. The financial investments in VQS are several times larger than those in vessel and gear, while the resource and entire sector is prone to heavy fluctuations. Consequently, a group is abstaining from financial obligations involved in VQS investments, and here the leasing market offers an alternative access. The use of the VQS was identified as access to the commercial fisheries, which pointed to the different practices and ways to organize that access. The exchange situations above were rooted in a multiplicity of practices. Age, health, behavior of colleagues, increase in regulations, retirement plans, and so on were important factors in determining the fishers' choices and actions in regard to selling and buying quota.

The material highlighted a diversity of fishing operations and indicated that market fluctuations in value had a significant impact on fishers' operational and financial decisions and long-term planning. Where the previous chapter looked at the overall development of the industry as a sum of actions (dismissing cultural heterogeneity), this chapter has contributed to the understanding of quota trade from the perspective of the actual people who are involved in this trade—both directly and indirectly—as experts of exchange. To do this, I found inspiration in Marx's work on commodity as having both an exchange-value and a use-value. To fully explain the quota trade, I examined the underlying value of fishing rights. These appear and are talked about as commodities, but to explain their value a better approach was found through the rent theory. The value of fishing rights comes from exclusive ownership and the fact that VQS can be monopolized by capital. The monopoly situation means that when the resource is put under exploitation by capital a rent can be extracted. The rent is in turn appropriated by the owner, whether through leasing or accumulation in the increased boat share. However, it would not be possible to capture any value or rent without the labor process, which underlines the social origin of rent in the concrete ways in which production is organized. In the next chapter, I will examine five such different ways fishing operations can be run and how the VQS is applied in actual use.

References

Dupont, D. P., K. J. Fox, D. V. Gordon, and R. Q. Grafton. 2005. Profit and price effects of multispecies individual transferable quotas. *Journal of agricultural economics* 56 (1): 31.

Einarsson, Níels. 2011. Culture, conflict and crises in the Icelandic fisheries: An anthropological study of people, policy and marine resources in the North Atlantic Arctic. Acta Universitatis Upsaliensis, Uppsala.

Grafton, R. Quentin. 1995. Rent capture in a rights-based fishery. *Journal of Environmental Economics and Management* 28 (1): 48–67. doi:http://dx.doi.org/10.1006/jeem.1995.1004.

Grafton, R. Quentin. 1996. Individual transferable quotas: Theory and practice. *Fisheries Reviews in Fish Biology and Fisheries* 6 (1): 5–20.

Højrup, Thomas, and Klaus Schriewer. 2012. *European fisheries at a tipping point*. (Estudios Europeos, No 1). Murcia: Edit.um.

Marx, Karl. 1976. *Capital: A critique of political economy*. Harmondsworth: Penguin.
Marx, Karl. 1981. *Capital: A critique of political economy*, vol. 3. Harmondsworth: Penguin in association with New Left Review.
Ramirez, Miguel D. 2009. Marx's theory of ground rent: A critical assessment. *Contributions to Political Economy* 28 (1): 71–91.
Schou, Mogens. 2010. Sharing the wealth. *SAMUDRA* 55:18–23.
Squires, Dale, James Kirkley, and Clement A. Tisdell. 1995. Individual transferable quotas as a fisheries management tool. *Reviews in Fisheries Science* 3 (2): 141–169.

Chapter 5
Access and Fishing Activities

Abstract In this chapter, I look at the implications of transferable quotas on the organization of production; that is, how fishing activities are structured around access to the individual and transferable quotas and how, in turn, the quotas structure the production. Therefore, this chapter will give a detailed ethnographic description of five different fishing operations and then compare them on a number of different fronts. This will direct us to some general differences in their modes of operation in relation to the vessel quota share (VQS) system and lead us to the next chapter, where the principal implications of the VQS for different *modes of production* will be discussed.

Keywords Modes of production · Large-scale fisheries · Small-scale fisheries · Social organization · Quota investment

The use of vessel quota share (VQS), in other words the use-value aspect, was briefly discussed in the previous chapter (Chap. 4, "The Commodity and its Exchange"). There, I argued that the use-value of the VQS commodity was the access to the commercial fishery and the right to transform a certain amount of those usevalues into a fish commodity for a fish commodity market. Through ethnographic material, I showed that this peculiar use-value is measured and talked about in kilos, and that the VQS use-value each year consists of a fluctuating number of kilos, a special feature of the VQS as a commodity. The VQS is talked about and treated by fishers as a commodity, a necessary means of production; but in the previous chapter, I showed that in important aspects it functions more similarly to land—giving rise to monopoly situations. The VQS was created through state intervention as a political, administrative, and management tool and commodity. It only appears as "fish" and as a commodity when talked about, exchanged and used. In that regard, the use-value expresses the right to participate in the commercial fishery, which also explains why it is sometimes referred to as "rights."

Heterogeneity

Use-values understood as a commodity's capacity to fulfill human desires are subjective and heterogeneous, and so are the uses of the VQS. Even for the same commodity there can be different use-values. But in order to fully understand the use of the VQS and its implications, we have to focus on the mode of production that transforms the VQS from a right into a fish commodity that can be sold in a fish market. In the simplest fishing operation, one person can run his or her own operation alone, perhaps with some administrative help from a spouse or children. This is not at all uncommon today in Denmark, where even rather large vessels can be operated by a single person. On the other end of the spectrum, there are companies that have several vessels, running with hired skippers and rotating crews as well as an administrative office on land.

Similar to other sectors, the organization of production, or the mode of operation, is rather diverse in the fishing sector. What this chapter will show is that the phenomena of leasing and quota pools make it even more diverse. Leasing out and in have become more or less fully integrated into the economic strategies of fishers. An operation that is 100% dependent on leasing will have to adapt to the conditions on the leasing market, and this type of fishing operation is a new phenomenon and distinct to market-based fisheries. The organization of production is full of heterogeneity, which I am attentive to, while also searching for common traits in the material. Not only can production be organized differently but the commodity delivered to the fish buyer can also be of a different quality and type. Some companies deliver frozen blocks of fish while others sell some of their catch directly to the consumer. In the following section, I will give an account of five different companies and how their VQS is transformed into landed fish. The five modes of operation have been chosen to depict the plurality of fishing operations and are, summarized:

- A one-man operation
- A young new entrant
- An operation based entirely on leasing
- A new company
- A large-scale operation

The One-Man Operation

Claus is the skipper and only fisher on board his boat. This one-man operation has its home port in Svaneke, one of the most eastern harbors of Denmark located in the eastern Baltic catch area (see Fig. 5.1). His boat is just less than 12 m long and is mostly engaged in trawling combined with some longline fishing in the autumn if the fish are widely dispersed. In the years before the VQS system was introduced, he was fishing under a coastal fisher arrangement for the Baltic Sea. Under this special regulation, which was only applied to the Baltic Sea, he was allowed to

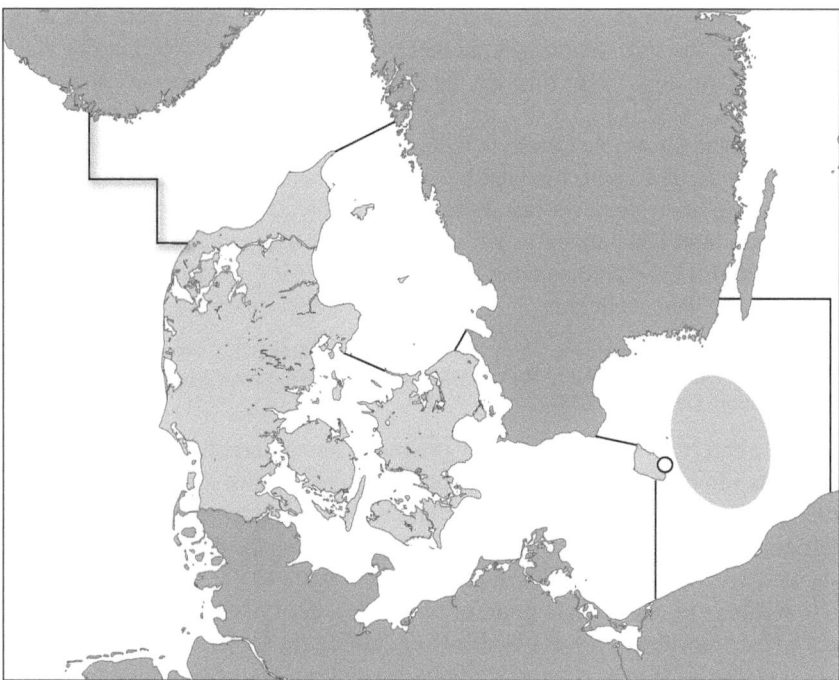

Fig. 5.1 Map showing the location of main fishing activities. The one-man skipper always lands in his home port, and trips never last more than a day

catch 40 t of cod each year plus some other species[1]. But for Claus the amount was too small to make a living, so in 2006 he moved with his family and fishing boat to Thyborøn, a fishing port at the opposite end of the country next to the North Sea. A year 2006 was critical for the VQS system, as it was just after the reference years used to determine the catch history (2003–2005) and the last year before the VQS system was introduced. Because of this his catch history reflected his fishery in the Baltic Sea and not the new planned activities in the North Sea. At the same time he could see the potential of acquiring more "fish" and making a living in the Baltic, a region to which he still felt attached. The alternative would have been to swap his Baltic "fish" for North Sea "fish." After only 5 months in Thyborøn, he and his family decided to move back:

> We were up there for five months, to see if it was viable. But then at the same time the VQS came, and I started to doubt, is that where I am going to be or not.... Because all of a sudden I could not get any quota up there. (Personal conversation, December 2011) [2]

---

[1] In that way it was a system of individual and annual but not transferable quotas. It is my impression that the system was popular but that the amounts were too small (Personal conversation, December 2011).

[2] With the change from the previous ration-based system to the VQS, he would not be allowed to land any North Sea fish; but in 2006 he could get a ration on the same terms as anyone else.

So Claus chose Svaneke in the eastern Baltic as his future home port, and in order to supplement his VQS he bought another vessel with VQS for cod in the Western Baltic, which he now swaps through the quota pool so he can stay at "home" in the Eastern Baltic[3]. Cod is by far the main species for Claus; and, because of the rising cod TAC (Total Allowable Catch) in recent years, he now receives an annual allocation of around 100 t. With his boat he can bring back around a ton of fish from a good trip, and the trips never last more than one day. It is "in and out" he tells me. His aim is to land 120 t annually, which for the time being would force him to lease some quota. But his high dependency on a single species, the vulnerable Baltic cod, is also one of Claus' concerns:

> Where I am most vulnerable is if they reduce our quota completely. Then you are in trouble, and you are locked in the Baltic Sea and you cannot get days at sea anywhere else. (Personal conversation, December 2011)

The Baltic cod stocks have previously proven to be vulnerable not only to overfishing but also to changes in the water condition and salinity levels. In other words, for Claus the VQS system reduces the flexibility to change gear, target species and catch area. This risk is partly reduced, Claus tells me, by the leasing system. But if the cod TAC is reduced, he will have to lease more to keep the same volume of catch. Claus is in many ways grateful for the leasing system. When he injured his waist a few years ago and was not able to go fishing for a whole year, he managed to make it through the year economically by leasing out his VQS. Being one of those who chose to engage in the VQS system and buy further VQS, he is quite happy with the VQS system:

> There are only advantages with this system, as long as you have something. For those without anything it is not as fun. And those who have something, they have the debt as well, they are taking a risk. The others could just as well buy and take that risk. (Personal conversation, December 2011)

In many ways, it seems Claus has found a balance between his VQS and the size of his vessel. He is only active in the Eastern Baltic and only embarks on one-day trips. While there are still clearly a lot of regulations that are an annoyance, through a little give and take he has found his place in the system. He has used the possibilities of leasing both in and out and also taken the risks associated with buying another vessel. His ambitions for the future are to keep his one-man business running and when he gets older to downscale production but keep on fishing as long as he can.

> I hope that when I get old, I too can go down and buy some fish from a fisher. That would be nice, to help him by shoveling some ice, or whatever you can do when you get old. It would also be nice to help a young guy if you could see he had the enterprise. (Personal conversation, December 2011)

Claus' plans for the future do not involve buying further VQS. More likely, he will sell when he reaches a certain age and continue fishing on leased fish or the VQS he has left. In this way, he is a good example of the many other operations that have purchased one or two vessels with VQS and supplemented this with leasing when

---

[3] Even though he is not more than an hour's steam from the Western Baltic.

needed. Those vessels that are not members of the VQS-pool system are likely similar to Claus and have adjusted their vessel, VQS, crew size and operation to achieve some sort of balance (Personal conversation, December 2011).

There are many factors to manipulate. Claus bought a vessel (that he never saw) and took the VQS from that, and with an increasing TAC—and a stable number of days at sea—he focused on trawling, instead of hooks and gillnets. When it has been a good day on the sea, he has one or two employees to help him gut the fish after he has returned; but the fishing operation out on the sea he can handle alone. The reason why Claus shifted from gillnets to trawl is also related to the VQS system. In the early months of the year the conditions for gillnetting are not ideal because of weather conditions, while trawlers are easily able to fish. This meant that when Claus began his gillnet fishery in March or April, the trawl companies had finished their own quotas and were busy leasing in extra allocations. In the first years of the VQS system, the TAC was quite low so most people were looking for extra allocations, including Claus. This made the leasing prices rise month by month. Since Claus would not lease quota in advance, when the prices were still low, he decided to sell his boat and buy a small trawler. In that way, he could finish his own quota and lease in before the prices were pushed up too much.

Direct Sales and Tourists

Claus is fishing from a town very popular with tourists, and he often finds himself surrounded by tourists or curious locals when landing his catch. The tourists, as well as the locals, regularly ask if they can buy fish from him, and he has made a website for interested buyers. There they can follow him on a webcam, see his position on a map, know when he will be back in the harbor and, of course, get the crucial information on what type of fish he is bringing in for sale. The first website was created by him and his daughter, with help from a local photographer. But now, with the support of an organization, the website has been professionalized and opened up to fishers from all over the country. It is of course optional to be a member, but according to Claus:

> …the advantages to being a member are straightforward and logical, you can with a minimum of effort easily triple your revenue, so when others hear about that, more will join.[4]

So even though he sells a limited amount of fish directly at the quay, he gets a much higher price than from the fish buyer. Baltic Cod, being small and mild in taste, is a low value product and does not bring in the highest prices when sold to the fish buying company. So, embarking in direct sales can bring the kilo price from as little as 5 DKR to 40–50 DKR. Thus, for Claus the direct sale contributes to increasing the income from his quota, because he can obtain a higher average price per kilo. He is in a way creating a short supply chain by bypassing the distribution links

---

[4] http://www.fiskerforum.dk/erhvervsnyt/print.asp?mode=erhverv&id=3586 (Accessed July 10, 2012).

between producer and consumer—thereby appropriating a larger share of the social value—and perhaps even adding value through the immaterial use-values created for the consumers: buying fish directly at the harbor, "talking to the fisher," etc. I will return to this discussion and perspective later. Since the first interview, he has also made deals with local restaurants to deliver filleted fish, which also brings in a better price.

Young New Entrant

The many—primarily economic—challenges for new entrants are also one of the main issues in the critique of market-based fisheries management. Not only do they illuminate the doubtful ethics of gifting a limited group of people—the initial right holders—a privileged ownership over a marine resource. The problem also entails the high entrance costs for the following generation, who have to invest not only in fishing vessel and gear, like in the "old days," but also in fishing rights. The following ethnographic case is an example of a young man who started his skipper career a few years after the VQS system was introduced. I first met this new young entrant while driving down the west coast of Jutland in March 2011, where my supervisor and I chatted with him while he was loading gillnets onto his vessel. Later, in January 2012, I met with him at his home where he lives with his parents. When I passed through the garage, I met his older colleague in the process of fixing some gillnets. We talked about the old days—how many boats there used to be and how few there are now—a very common theme when you work with this topic and visit different fishing towns around the country. I left the older colleague preparing gillnets and entered the kitchen to begin the interview. I was interested in his point of view and operation, as one of the few young people I have encountered in the fishing industry. Not surprisingly he is from a local fishing family:

> Yes I grew up here and my dad used to be a fisher, but it was never the plan that I should become a fisher. But then, what were the future options back then, there was nothing and there were no other jobs to take. So I went out sailing and just continued to do that. Then I could just as well become a fisher. I really enjoyed it then. I think it is a wonderful occupation, the open air and all that. (Personal conversation, January 2012)

The career choice was neither intended nor does it seem very surprising. He grew up in a fishing family in a remote town where the choices were limited. Either one moved away for a job, an education or one stayed and had to choose between the few work options available in the town. After trying out fishing, he soon wanted to be the skipper of his own boat and inevitably stood face-to-face with the VQS system. As a new entrant of less than 40 years old, there is the possibility of acquiring the status of a "first time established" (which is newly established for the first time). With this status, you receive a portion of VQS as a time-limited loan. The loan of VQS is administratively determined and equals roughly a value of 20 % of the average turnover for a vessel of the same size. In other words, the amount is not enough to make a living. The VQS loan ends after 8 years but is reduced by a quarter each year after the fourth year, giving the new entrant a smoother start but also a growing gap to fill by leasing or buying VQS[5]. In this case the gap is filled by a mix of strategies, and leasing from the formal pool system is just one of them:

> I am member of the pool and lease some cod. I do that once in a while, but we do not lease much. This year [2011] we have leased fish for 50,000 it is not more than that. The rest is something we have gotten cheaper from all kinds of other people. I know a lot of people here. Sole, they have not caught them here at the end of the year [...] then we lease it privately from another fisher at a cheaper price. Then we talk with the different people, I kind of know if they have not fished it. (Personal conversation, January 2012)

What is interesting here is that he differs between leasing in the formal and web-based pool system from leasing from VQS holders he meets and knows. Technically the paperwork for both will have to be dealt with by the quota pool, and both types of leasing will appear formally as leasing in the pool system, like all other leasing transfers. Throughout the interview he refers to "leasing" primarily as the formal procedure in the pool system. In other words, he does not use the word when describing the deals he makes locally, separating these from the administrative discourse on leasing. So, through his network he gets favorable offers that contribute to his annual allocation and production. In return, the individual leasing out can be sure that his VQS will be leased independent of the price developments in the formal leasing market. Even though there is money involved, they describe this as helping each other out locally.

Fishing Strategy

On top of the time-limited VQS loan, the leased fish and the fish he 'gets' through his social network, he fishes VQS that belongs to his father's boat, mainly of cod and turbot. His annual landings are thus made up of several different sources of VQS. Some, like the leasing, is subject to fluctuations in prices and seasonal changes, while the VQS from his father and the state loan for first-time established is more stable throughout the year. It is interesting to note that securing VQS is a constant task this new entrant has to engage in to have enough VQS for his operation throughout the year. One way to solve this would be to lease all the required fish in the beginning of the year, which would make him less dependent on the leasing market and enable him to better plan the year in advance. This is the practice for some:

> That is why people are leasing right now at the beginning of the year. I have never done that, I have leased when I needed it. It might sometimes have been a mistake. But I have not been in lack of anything—and I have never had problems leasing when I had to. (Personal conversation, January 2012)

Leasing at the beginning of the year also means running the risk of not being able to catch what is leased, a situation he is trying to avoid. Indeed, the beginning of

---

[5] In the 8 year period the restrictions on buying small shares of VQS are looser, which should make it easier for new entrants to by smaller amounts of VQS if these are for sale.

the year does see a lot of activity on the leasing market, where VQS holders seek to make up for the gaps in their VQS-holdings, in order to plan their fishing activities with greater certainty. But this is not the practice in this case. Another factor that explains his procedure is that the leased fish has to be paid up front, and therefore cash or credit is needed to lease fish—with the inherent risk of not catching it. Instead, he leases VQS bit by bit when he needs it, a strategy that has been working so far, but which he also knows means he sometimes leases at a higher price than if he had leased the VQS at the beginning of the year.

The Vessel and the Fishing Activities

In 2009 the new entrant bought his boat, an old vessel built in 1961, for the price of 220,000 DKR. The vessel had been stripped of VQS[6], which explains its low price and why he is dependent on leasing and the state loan of VQS. The vessel is 14 m long and just short of 20 gross t. The fishing activities take place in both Skagerrak and the North Sea and as far down as in the English Channel. He moves around in catch areas and places in order to be where the conditions are most lucrative, and his niche changes according to prices and weather conditions.

> Last year we fished for around 1.2–1.3 million [...] We caught a lot of turbot and sole, that is what we made the annual income on, turbot and sole. We only got 12 tons of cod because of all the wind. (Personal conversation, January 2012)

Being flexible means changing between gear, area and target species, and not always fishing from the home port. Most of the fishing trips last between 1 day and 3–4 days; but some of the trips, such as those for Dover sole, take place in the English Channel off Holland, and the fishing trips there last 7–8 days (see Fig. 5.2). On board he is the skipper and there is only one other fisher besides him. The crewman is an experienced fisher who was unemployed. To try something different, he had been working for the windmill producer *Vestas*; but his job was made redundant, and he was fired just after the financial crisis in 2008. After being unemployed for some time, he called and asked if he could join in the fishing. Even though the young man is the skipper and therefore the boss, they work as a team:

> We take as much as we can do and set the nets according to what we can cope with. Sometimes we overwork, he is skilled. (Personal conversation, January 2012)

Since a great part of the fish is leased in, either through local contacts or through the quota pool, the crewman has to pay half the leasing cost. Most commonly the leasing cost is deducted from the income from the sale. After the deductions (oil, ice, leasing, etc.), the boat share gets 70 % and the crew share is 30 %. Apparently, as the owner, he does not differentiate between the boat share and his own share:

---

[6] In 2001 it was traded for 700,000 DKR and then again in November 2007 for 4,550,000 DKR before it was sold without VQS.

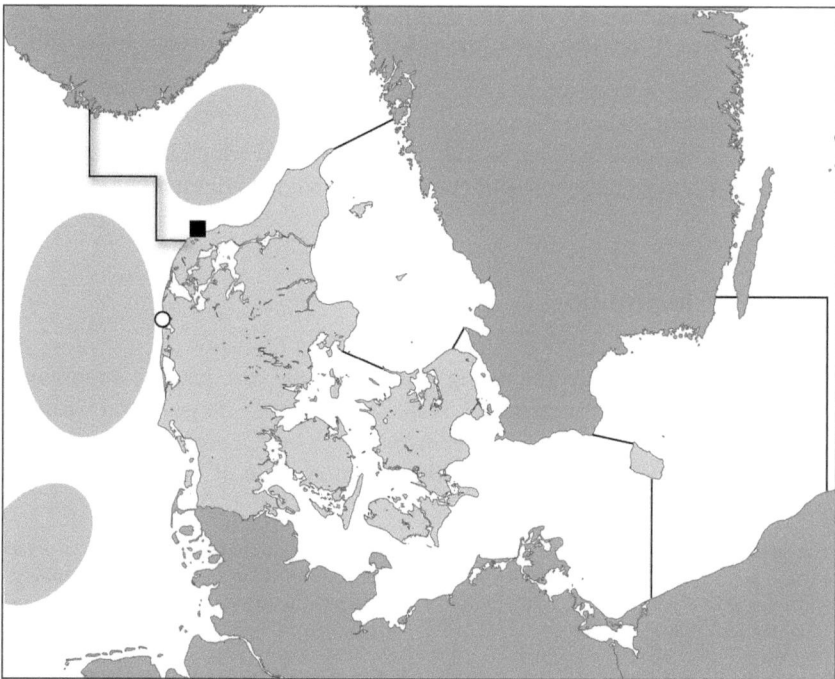

Fig. 5.2 Map showing locations of main fishing activities. This operation sometimes lands and operates from Hanstholm (*black box*) when the price conditions are better there

> The ship and my salary runs over one account, I live at home and do not use any money anyway. I take out some money if I need it—that is how it is. (Personal conversation, January 2012)

So the young new entrant lives at home and, according to himself, is not a big spender. To have a chance in the VQS system he needs to keep his costs down, and there is one peculiar method he uses to keep the costs of his operation low. He spends time repairing and preparing his gillnets himself, a time consuming task that, accordingly, most others would pay someone else to do.

> That is to put new nets in, it takes some time. A lot of spare time is consumed by that. Even during the days when I am at home, I am up at seven and work from eight in the morning to five in the afternoon. I take two or three breaks, but otherwise it is out there [points to the garage] it takes place or down at the harbor. It is like a normal working day. (Personal conversation, January 2012)

It is interesting to note that he refers to the time he spends preparing nets both as his spare time and as a normal working day. The translation from the Danish word "fritid" could perhaps have been "free time," which would then be a more precise description of the contrast to the "bound time" on the vessel, and not to the time not working. I will return to these aspects of time and *free* time both in this and the next chapter. The gillnet production is only for his own operation, but it allows him to use more gillnets at a reduced cost.

…last year we had 200 turbot nets, the year before we had only just 120, and this year we will have 350, we are focusing on the expensive fish. (Personal conversation, January 2012)

Gillnetting flatfish is labor-intensive and tough work, but the high-quality fish product from gillnetting yields a high price in the fish auction. The operation is further specializing in lucrative species such as sole and turbot, which are harder to catch by trawl and, when caught with gillnets, retain the same high quality.

Thrift and Flexibility

In the stormy winter months the young new entrant has also taken a wage-working job for the coast guard, in emergency response. This way he gets paid while he is at home preparing nets. All in all, it seems that he uses economic cautiousness and thrift as a way to reduce the costs in his operation. One of the objectives of this strategy is to avoid taking bank loans:

> The ship I have was bought and paid the day I bought it, I paid in cash. I have never owed anyone as much as one krone, and if you do like that you can always get a business running, you are stupid if you cannot do that, to fish enough to pay the insurance and such. (Personal conversation, January 2012)

For now, thrift and low costs are enough for the new entrant to avoid credit institutions, and he has built up enough cash to pay for the leasing of VQS, which he does bit by bit and not at the beginning of the year. However, in the near future the VQS loan will be withdrawn, though he might be able to buy his father's VQS at "cheaper money." He is reluctant to invest in VQS, which he sees as unstable, much like stocks on a financial market (see Chap. 4). At the end of the day, he informs me, he would rather find a job on land than be dependent on banks.

The Tenant

The following ethnographic account concerns the same person and his involvement in four different fishing operations, four different ways to organize the use-value of the VQS. Since the 1980s, "the tenant" has been fishing from Gilleleje, one of the major ports in Denmark and the largest fishing port on Zealand (see Fig. 5.3). Just before the institution of VQS system, he was the owner and skipper of an operation that focused on Norwegian lobster and cod trawling. Their catch history resulted in a VQS allocation that was very much concentrated in Kattegat, with a small amount in Skagerrak and the Western Baltic Sea. At that time there was one other person on the boat, and they were fishing as share fishers: the boat received a 52 % share and the two fishers shared the remaining 48 %. After the VQS allocation they continued more or less in the same pattern as before.

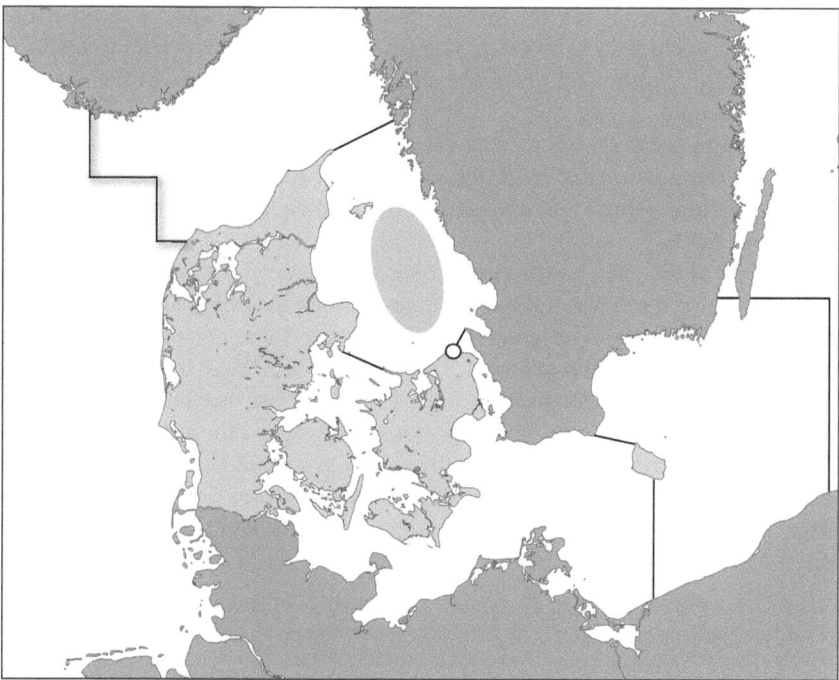

Fig. 5.3 Map showing location of main fishing activities; mainly lobster trawling based entirely on leased fishing quota

> We continued almost in the same way. We fished most of the year [after lobster], and then a couple months or three each year we fished for cod. For that we were also down in the Baltic Sea, a few trips, otherwise we fished here at home. (Personal conversation, December 2011)

One year he leased extra lobster because of extraordinarily good catches, but he did not deduct the cost from the crew shares. Even though he was aware of the collective agreement between boat owners and crew, which stated that the cost of leasing could legally be deducted beforehand, he did not do so, but paid for it from the boat's share:

> Yes, I did, because we were only two and he had been a part of it so many years, but I think it was in the collective agreement, that it [the leasing cost] could be deducted in advance. It can be deducted before the shares. [...] I also leased some out, should that income then had been added to the boats fishery, and included in the crew shares? I am not sure. (Personal conversation, December 2011)

Because of loyalty and perhaps because of the small amounts, he did not make use of his legal right to deduct the cost of leasing. But that has changed now. After a few years a set of new and stricter conservation regulations were introduced. These launched a number of closed areas which covered the main grounds where he had been fishing in Kattegat. On top of this came not only the "days at sea" regulation, which allocated less fishing days to the operation than they needed, but also the future prospect of renewing the motor—a considerable expense. Fed up with the regulations and reluctant to invest in a new motor, he decided to sell his vessel with VQS. He had just turned 55 years old and could now make use of the special tax scheme for people closing down a company for retirement. After selling the boat he continued to work on his former vessel for 3 months, now as a hired skipper working for the new owner. With the change in ownership came a change in the payment structure, which was changed so the boat share was now 58% instead of the previous 52%.

At the end of the year, he stopped fishing and searched for employment on land. However, finding a job was not easy and he returned to fishing again. He then took a job as a hired skipper—on a vessel owned by the same person who had bought his former vessel. Now the payment structure was a 60% share to the boat and a 40% share between the people working on board. After that season he bought an old vessel without VQS from—perhaps not surprisingly—the same individual who had bought his former vessel and employed him as a hired skipper. Without a VQS allocation, he is now basing his operation entirely on leased fish, mainly Norwegian lobster. Because of a large TAC on Norwegian lobster, the leasing prices are also relatively low:

> They are not that expensive, at the moment I think they are traded at 1.5–2 kroner per kilo. That is the rent, and you get 60–70–80 kroner sometimes up to 100 kroner for them [at the auction]. So it should be possible to make a business fishing on leased [lobster], actually a lot of people are doing that, several people do not have that many lobster, those that fished cod back then, they did not get that many lobster. (Personal conversation, December 2011)[7]

For the VQS holder with excess Norwegian lobster quota, it is better to get 2 or 3 kroner per kilo than to get nothing at all, so the plentiful supply means the leasing cost is quite low for Norwegian lobster. At the same time the Norwegian lobster is a high value product, and the auction prices can be as high as 100 kroner per kilo, which creates a significant difference between leasing price and landed value. It is this difference that the tenant uses to be able to run a fishing operation without having any fishing rights at all.

The Vessel and the Future

Because he was previously an active skipper and because of his age, there is no new entrant loan of VQS from the state to help him get started, and he is not planning to invest in either VQS or a new vessel. He bought his new vessel for 600,000 DKR, which included some "days at sea" rights that, according to him, roughly account for half of the value of the boat. It is an older 15-m trawler from 1983, but he has no

---

[7] He is referring to the decline in the cod TAC that means the number of use-values of cod VQS is lower; and hence a large group of fishers are now leasing Norwegian lobster to fill the gap from the missing cod.

intentions of buying a new one. Instead, he is satisfied with the limited investments he had to make:

> This is not such a big investment. Of course, there are costs for insurance when you have a boat, and it needs maintenance while not in use, but it is not as big an investment as if I had bought a boat with fish and everything. […] If it all fails it can just stay here [in the harbor], if I find some other job and have to spend too much time on that. (Personal conversation, December 2011)

Instead of investing in a new vessel and VQS, he bought a cheap boat without any fishing rights and now depends 100% on leasing Norwegian lobster. While the leasing price of Norwegian lobster seems to be rather stable, there is very limited flexibility for the tenant to switch to another fishery if the Norwegian lobster fishery or auction price should fail. If he based the operation on other species, the span between leased cost and landed value would be much smaller and leave less to be shared. Instead, in this case the small investment leaves him with plenty of flexibility if the right job on land should show up. So, the specific flexibility that this mode of operation offers is between fishing as an occupation and the ability to shift to another job if the possibility should arise. The vessel more or less keeps its value because of the kilowatt rights attached to it, and thus he only needs to keep it afloat and capable of fishing until the day he decides to sell it.

In a short number of years, my interviewee has been part of four significantly different operations. First, he was the skipper on his own boat with a good portion of VQS to base his and his partner's fishing on. After selling he was employed as a hired skipper, first on his former boat and afterwards using another person's vessel and VQS. Finally, he was a vessel-owning skipper again, but this time dependent on leasing in the VQS that formed the basis of his fishing operation. It could be tempting to ask if, in hindsight, he would have chosen a different strategy. He could have used his allocated VQS as collateral to take a loan, and with that bought a brand new vessel.

> It would take many boats to pay off the new one. Our fishery will not be able to pay of a new one. Then I would need to buy a few more, and that is a vicious spiral, I would have to buy a few more with the same amount of fish to make it viable. I was not interested in doing that…but some have done it. (Personal conversation, December 2011)

His current mode of operation is designed specifically to avoid extensive investment in VQS and the following expansion of fishing activities, and it leaves him with little to no risk if conditions should change. In this way, the tenant's situation is in clear contrast to the following case of an expansive company.

The New Company

The fourth fishing operation is owned by two brothers, who decided to join forces when the VQS system was introduced in 2007. Before that they were both fishing on a boat owned by one of the brothers, and they had been doing so for 15–20 years.

But when the VQS system was about to be introduced they decided to join forces in the new system and form a new joint company. Together they bought a boat[8] from a local fisher and took the VQS from that into the new company. The brother who had previously owned a boat sold it with its VQS to their new joint company and subsequently bought another vessel on the side, which he has as an independent company. The two brothers lease VQS from this independent company to their joint company. Partly due to the fact that they were now two owners, and partly because of new accessibility of financial support, the two brothers invested not only in the vessels mentioned above but also in a new vessel, which was ready for operation in 2009. The fact that their VQS holdings could be used as collateral enabled them to establish a new company structure and to invest in further fishing rights:

> Things started to happen, at that time you could sell hot air, you could get loans in the bank, because the fisheries got such a tailwind with the VQS system that you could use your fish as security, so you could get a loan [...] Now you could go and put your fish as security, and you could buy fish from your colleague as part of the system. (Personal conversation, December 2011)

Therefore, a function of the VQS was to serve as collateral for bank loans that enabled the two brothers to form a company, to build a new vessel and to expand their ownership of VQS.

The Vessel and Fishing Activities

The new vessel was constructed to fit the brothers' ambitions in the new management system and to concentrate on the consumption fishery. This was a shift from their earlier activities focused on both consumption and small pelagic fish. The new boat was an investment of 12.5 million DKR, and it substituted a vessel from 1959 that was a little longer (18 m) than the new one but less than half the size measured in gross tonnage (47 and 110 GT). In order to be able to trawl in the area near their home port, the vessel was constructed to be less than 17 m, which is the maximum length allowed for vessels trawling in the so-called area 22 (also called the inner waters catch areas). The activities on board are organized based on a four or five person crew[9] working in a rotation system:

> [...] three are out sailing and one man at home, three weeks away sailing and one week at home. Then it goes on like that. We have been doing so for two years.

The people on board can be divided in two groups: the two brothers who are the owners and the crew. One of the owners will always be on the bridge functioning

---

[8] This ship was from 1986, and its value at that time was 1,475,000 DKR. In 2006 they bought it for 1,800,000 DKR, and after stripping it for rights (VQS, tonnage, kilowatts) it was sold as a "retired fishing vessel" for 30,000 DKR.

[9] At the time of the interview, December 2011, they were four but planning to hire one more crewmember for the coming year.

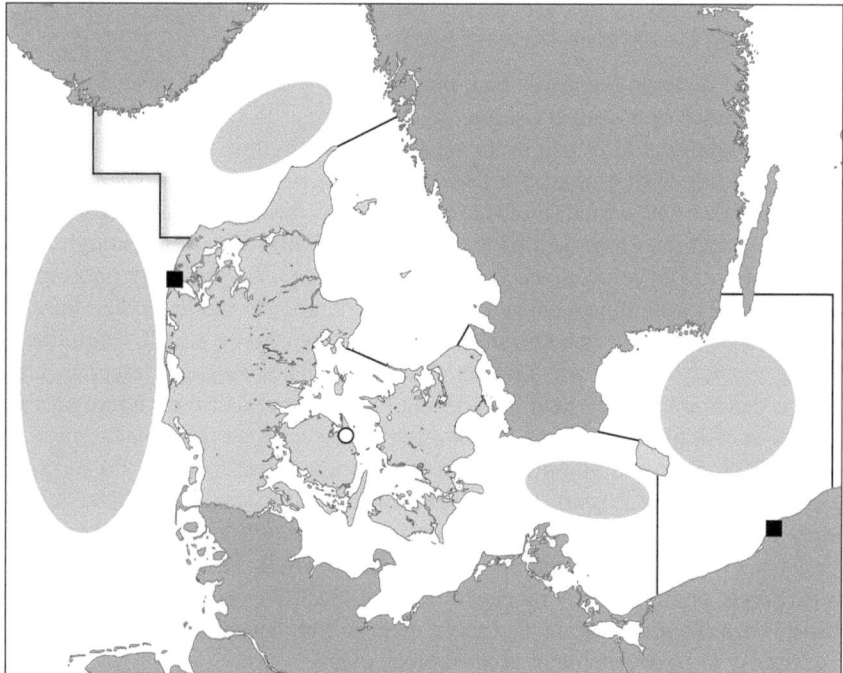

Fig. 5.4 Map showing locations of main fishing activities and landing places (*black boxes*)

as skipper, with a crew working on deck. When the two brothers are on board at the same time, one of them (the younger brother) works as crew on the deck while the other functions as the skipper. The crew members are all of Polish nationality and employed through a recruitment agency. The crew is paid partly by a percentage—sharing the income from the landings with the two owners—and partly as a fixed salary, the latter being rather unusual in the fishing sector. According to the skipper interviewed, this pay structure is based on the crew's preference, as they are guaranteed an income even in rough weather when the operation is on standby.

Production and Landings

Even though the new vessel was constructed so it would be permitted to fish in area 22, most of the fishing activities take place away from the home port in Kerteminde and sometimes in distant catch areas (see Fig. 5.4). Over the course of a year the vessel is hardly ever in its home port, except for a few short periods:

> That is when we are at home changing gear. We have a gear shed here and one gear shed in Strandby, because we spend quite some time in Northern Jutland fishing Norwegian Lobster. With the new vessel we have not been fishing for more than two days just here outside

Kerteminde. Usually, we have been further down in the Baltic Sea. There is more cod down there. (Personal conversation, December 2011)

When further down in the Baltic Sea, the company lands their catches in Poland to the Polish branch of a Danish fish buying and processing company. Not only is the price a bit higher per kilo than in the nearest Danish harbor, but this is also a more convenient place to disembark for the Polish crew. Because of the specific regulation in the Baltic Sea, which (in 2012) automatically gave each participant 160 days at sea, they spend a great deal of the year in the Baltic Sea, with enough days at sea to lease in extra allocations. However, throughout the year they participate in seasonal fisheries in almost every corner of Denmark. To facilitate this, they have a gear shed in Strandby in addition to the one in their home port Kerteminde. The company is thus in many ways widely distributed. They target different species in different catch areas, they land in different harbors and visit their home port rarely—only to change fishing gear. In order to run the vessel at full capacity they have to lease VQS. Between 30 and 50% of their landings are from leased VQS, with the cod fishery in the Eastern Baltic especially based on leasing in VQS. The VQS system enabled them to invest in both a new vessel and in further fishing rights, which made their total investments over 22 million DKR. The vessel is out for fishing whenever possible. They managed through the VQS system to establish a new company, but the investments also force them to run the vessel almost nonstop for the company to be profitable or at least break even:

> When you have such a ship, where you owe that much money away, then the vessel needs to be out fishing at all times. (Personal conversation, December 2011)

But it is not only the financial cost that puts economic pressure on the company, it is also oil prices and leasing costs. In order to be independent of leasing they would, according to the owner, need to invest another 20 million DKR. So even though the two brothers used the new exchange-value to reshape their company and expand, they are still small compared to the "big guys" in the sector. Faced with the new system they chose to react and be progressive, but they are also now in a situation where the financial stress is significant:

> It is not enough to make a turnover, you have to make profit. There is no guarantee you will have a large income if you have just as big investments. That is the central point in this system; you have to be a good businessman. (Personal conversation, December 2011)

Here the skipper outlines the exchange-value and use-value aspects of their practice; you have to be both a good businessman and fisher. Taking into account the large amount of investments (in vessel and VQS), it is crucial that they continue to serve the company, i.e. that the TAC is not reduced for their main species and loan conditions are favorable and stable. In that way they are highly dependent on and bound to the developments in the TAC regime and the financial climate:

> The future. I can tell you, it is only a question about survival. Because after we have had the financial crisis, it is only a matter if you can deliver a satisfactory result, if you cannot the bank will end the cooperation. Then you are left with nothing. It is cold and cynical. It is a knife at your throat, if you do not deliver good results. (Personal conversation, December 2011)

The Large-Scale Operation

The last ethnographic example in this chapter is a large-scale operation, its fishery mainly driven by a large trawler with its home port in the Eastern Baltic. In many ways, the VQS system has changed things for this operation. Before the introduction of VQS, the activities of the company were mainly focused in the North Sea, due to regulations that limited large vessels in their home waters; but today they have reestablished a large part of their fishery in the Baltic Sea. Two people own the company. One is the skipper, who takes care of the operation, while the other is only engaged through his investments in the company. The investor has a background in the fishing industry and is involved in a number of other companies in the region, though he is no longer involved directly in fishing activities. The ship was built in 2000 before the introduction of the VQS system, and in the first years the new boat was losing money:

> There were incredible costs related to the operation, we had to be on the sea all the time, and the cod, we were not allowed to catch it here at home. So it was unbelievably negative, the first five years we were only losing money. (Personal conversation, December 2011)

As we heard in the previous chapter, the two owners did research on market-based fisheries management, and, based on their research, they quickly decided to invest in vessels and quota to acquire more fishing rights:

> Then there were talks about this system. Then I read a lot of reports from Iceland, Scotland and Germany, which had introduced it before us, Individual Transferable Quotas, there were a lot of reports on it. I could see, and my partner could see, the possibilities in it. Either we would have to sell or we should go on and buy. Then we started to buy before it was decided, it was decided in October 2005, and at that time we had already bought the first two vessels. (Personal conversation, December 2011)

Since then, the company has bought 10 vessels in total, all with fishing rights for Baltic Sea cod. In total, the company owns VQS worth around 100 million DKR, a majority being the value of the VQS. This purchase demanded access to capital, and the two owners have taken joint loans at low rent. Because of the large value of the other companies owned by one of the co-owners, as well as the collateral in the fishing operation, the two owners can get an interest rate much lower than most would be able to. As the quote above reveals, based on their research the company started buying vessels in 2005, more than 2 years before the VQS system was properly in place. At one point, they had six boats running at the same time, and the fishing rights from these were transferred to the large trawler as soon as the new system was actually functioning. Because of the regulations that limited the fishery in the Baltic Sea, the catch history of the main vessel itself primarily reflected the fishery in the North Sea. Since then, they have bought vessels with fishing rights in the Baltic Sea to build up VQS for a whole season there. Out of the ten vessels the company has bought, nine were from the company's home port and one was from a neighboring town. Their investments have thus been focused on building up fishing rights in the Eastern Baltic Sea.

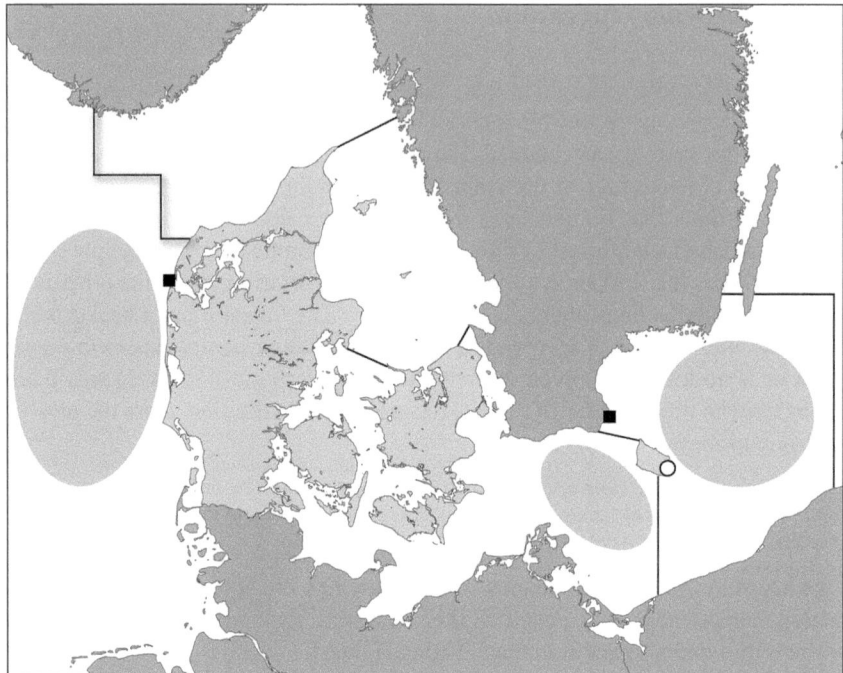

Fig. 5.5 Map showing locations of main fishing activities and landing places (*black boxes*)

Fishing Activity and Strategy

The company's main vessel is a large trawler originally built in 2000, rebuilt and enlarged in 2010. Today, it is just over 30 m in length and has modern cooling and handling facilities on board. According to the managing skipper, their strategy is to deliver the highest quality fish possible to the market, which is achieved through careful catch handling on board. Therefore, the crew is also larger than what would be technically needed to handle the fishing activities. The fishing activities are annually divided between the North Sea and the Baltic Sea (see Fig. 5.5):

> We leave for the North Sea around the 15th of April; that is when our season there starts. At that time the quality of the Baltic cod is so low that we are happy to switch. It becomes lean when it enters its spawning period. Then we are up there until the 1st of August and then here at home for the rest of the year […] Turbot, lemon sole, cod. When we are up there we land all our fish at the fish auction in Thyborøn. Then we take a break in July, in the industrial holiday season, where we are not running but taking holidays. It is much more relaxed now with the VQS system, we can better plan our fishing seasons. Before we had to fish all year round, we had to fish, because it was a ration-based fishery. (Personal conversation, December 2011)

The activities are split into two seasons. The longest season is "at home" in the Baltic Sea trawling for cod, a season that spans the period from August to the middle of April. After that the vessel steams to the North Sea and spends 3 months fishing different demersal species. Between the two seasons they have a fortnight break in July, following the main holiday season for schools and the industry. The fishing season in the North Sea is based on the vessel's historical catch and initial allocation of VQS, while the fishing for Baltic cod is mainly based on VQS from other vessels bought locally. They switch between the seasons at a time where the quality—and thus the price—of the Baltic cod is at its annual lowest. Because of the two very geographically different seasons the operation also has two different ways to deliver the landings. When fishing in the Baltic Sea the vessel lands its catches in Sweden. Here fish buyers pick up the fish by prior arrangement. Their average prices are higher than most others, partly due to careful catch handling and because they have built up a reputation for good quality:

> The handling of the catch on board, we are plenty of people, and we are almost down brushing their teeth. They are put in the boxes as straight as arrows entirely fresh caught. The buyers are asking for our fish, asking for when we will be landing fish. We can sell together with the others and get 20% more than they get per kilo, even though they were all caught in the same place. It is a reputation we have built up over several years. (Personal conversation, December 2011)

When fishing in the North Sea they always land at the auction in Thyborøn, and there they sell their fish to European fish buyers. The introduction of VQS has made the establishment of a reputation easier, since the deliveries are more stable and the vessel can now have a much longer season in the Eastern Baltic. Before the VQS system they were only allowed in the Eastern Baltic for 2 months. Now it is the VQS that determines the length of the season, in combination with a maximum of 160 days at sea.

Labor Organization

All in all, there are eight people taking part in the actual operation of the vessel. When the vessel is running at full capacity, six are on board while two are at home. When running at normal capacity, there are five on board and three at home. They all rotate to ensure that each person has one rotation at home followed by two or three rotations on board. A rotation period lasts for roughly 2 weeks. When the managing skipper is not on duty, one of the crew members—the same each time—takes the turn as skipper, and they arrange their turns between the two of them. The six other crew members work out a schedule for their rotations. While a few times each year the crew cannot agree on a schedule and the managing skipper has to plan it for them, the crew members mostly take care of the scheduling themselves. The people working on board get shares of 36% of the net income, while the remaining 64% goes to the operation of the vessel. The people involved in the operation and ownership are thus segmented in a number of ways. There are the two owners, who make decisions on the strategy and investments. The skipper and assisting skipper run the vessel, operation and plan their rotation; and the crew provide the manpower. The crew has nothing to do with the administrative handling of VQS, of buying and leasing fishing rights:

> They do not feel any big difference between this system and the previous, with buying of quotas and all that. They have no influence on it. They have confidence that I obtain enough fish. (Personal conversation, December 2011)

The future plan is that the assisting skipper will buy a portion of the company to facilitate a step-by-step takeover. With a value of 100 million DKR, it is necessary to have a conscious plan to hand over the company. The TAC for the Baltic Sea has been increased in recent years, which means that the company now has too many kilos of quota and therefore cannot catch all of its allocation. To solve that problem and to secure even more VQS for the company, they recently bought another, older trawler. The owner of that will keep on fishing with his old vessel, but now as a hired skipper with hired crew. If the TAC should decrease, the older trawler will be phased out and the VQS transferred to the larger, newer one.

The Great Expansion

After strategic research on the new challenges in the VQS system, the company engaged in an expansion and bought 10 vessels to obtain a full season in the Eastern Baltic. Combined with their initial allocation in the North Sea, this gave the company a base for a full year of fishing. They now run full-scale production with a modern vessel and significant investments. Good loan conditions have enabled the company to expand, and they have built up a reputation for high quality delivery. There are two owners, one of whom is the managing skipper, and on board they have a large crew who takes part in the production on shifts. One of their concerns is their long-term strategy; in other words, who will be able to take over the company. There is a concrete plan for the assisting skipper to take over a share, but the company anticipates that the ownership regulations will be changed so that joint stock companies can invest in fishing rights. In the long run they also expect the fish processing industry to be engaged in quota ownership.

Discussion: What Is New in the VQS System?

Above I have given accounts of five different fishing operations. They differ in many aspects and were, of course, partly chosen for the purpose of illustrating the variations in modes of operation. At the same time, they are not isolated or disconnected but represent some of the principal ways to structure operations that are found around the country and in different constellations, sizes, and numbers. The small one-person operation; the new company structure; the share-based operation;

Discussion: What Is New in the VQS System?

the large-scale operation; and the new generation of skippers exist side by side. By contrasting the five different operations described above, it is possible to see how many factors in a fishing operation can be manipulated and put together in different ways. Some of these are new and caused by the VQS system, while some are unchanged from earlier fishing practices and policy regimes.

It would be too simple just to call one small-scale and the other large-scale. There are numerous qualitative differences that would be missed in such a linear projection. They differ in how they maintain vessel and gear; how they think about and obtain VQS; their financial configurations; their human organization; their mobility; and how they plan fishing activities and fishing seasons. When catching fish they differ in gear types used and in the handling of the catch, and most of all they differ in their social organization. At the end of the fishing trip they have different strategies for selling the catch, different pay structures and ways of paying back and obtaining loans, and so forth. They differ in their annual, weekly, and daily practices, and also in how they perceive a career and make room for a new generation of fishers.

The varied modes of operation above also demonstrate that different meanings are attached to the practice of fishing. The unique structure of each operation represents a cultural practice linked to specific aims; ideas of what a career is; and understandings around being independent, in charge, hired, flexible, and so on. The central question for this inquiry is how the *VQS system* has changed all of this. In the following section, I discuss how the new *access through a market* influences fishing practices. When doing this, I attempt to explicate the general principles in the ethnographic material presented above. I look at how the practices described exist in relation to the VQS, with its limitations, risks, and potentials. What do the above descriptions look like if we put the "fish" in the center?

The small one-man operation invested in order to obtain an annual allocation that would secure him close to a full season in the Eastern Baltic. Since he actually invested in quota for the Western Baltic, he needs to swap quota with other fishers. Consequently, the dynamics of one TAC greatly influence his practice. If the TAC for the Eastern Baltic cod should fall, he will have to lease even more to maintain a full season. If it should fall dramatically he might be forced to take part in other fisheries entirely based on leasing. His dependency on leasing also made him shift to trawling instead of gillnets. He was reluctant to lease extra allocations before he had finished his own quota and cautious of the fluctuations in prices in the leasing market during the year. With the almost exclusive focus on one species, another issue is his dependency on the landed value of the catch. The price given by the fish buyer greatly impacts his ability to sustain his operation. To improve his resilience he is using direct sale as a strategy to be independent from fluctuations in the general market. By improving his income per kilo he needs to lease fewer kilos each year to maintain his operation.

The young and new entrant was not active when the system was introduced. As a consequence, he faced challenges in establishing a position in the system. He solved his need for VQS through a combination of several sources. As a "first time established" he obtained a temporary loan of VQS from the state and also had access to his father's VQS. On top of that he depends on leasing in VQS, which is done through a solid network and local contacts. The reluctance to invest in VQS forces this operation to be *mobile and flexible* and target different species, depending on when the VQS is *available* and *catches and auction prices are good*. In addition, the leasing market is bound up in other structures such as weather and markets prices, which in turn frame the conditions for the new entrant. Without substantial ownership of VQS, it is impossible to say for sure where, when, and for what he will be fishing, although the operation has established a focus on high-value flatfish. His comparative advantage is that he can land a high value product that demands a large amount of gillnets and hard manual labor. To be able to have that many gillnets and to keep the costs down, he spends his "free" time preparing nets, putting new nets onto the lead and floating lines. To obtain the VQS from his social network, he needs to maintain this network and his awareness of excess VQS among his colleagues; and to catch it he needs to be flexible and mobile.

The tenant was originally given VQS, but after selling his vessel he returned again to his fishing activities as a quota-less vessel owner. When he reentered as an independent skipper he did so with a low investment in an old vessel and a practice based entirely on leasing VQS. The operation is focused on a fishery where the difference between the leasing price and the auction price is significant enough to make this practice possible. He is of course very dependent on this condition to remain constant. If it does not, he is left with a bad business model. His main risk avoidance strategy is to stop fishing—to be able to quit without losses. Therefore his investments are as low as possible and he fishes from an old and almost run-down vessel that he has fixed himself. The operation is not designed to grow but to provide the platform for two or three people to engage in fishing activities through leasing.

The new company restructured in the new policy system by forming a proper company, investing in VQS and ordering a new vessel designed for their updated operation. Based on that and leasing, they have VQS to run the vessel more or less nonstop if the weather allows it. With the new company and new vessel they have hired a crew who is mainly paid by a fixed wage. Their own VQS is focused on the demersal trawling in the North Sea and the Baltic Sea. In addition, they also trawl Norwegian lobster. They have expanded and reshaped the operation through significant investments in VQS and a new vessel. In order to pay off these investments and to secure a profit from the operation they must maintain the vessel running at full-scale, which requires additional leasing of VQS. Thus, 30–50 % of the VQS in kilos are leased in. The economic situation is also expressed through the fact that they have a hired and wage paid crew in order to keep labor costs down. Because of the scale of the operation and the type of fishing they are engaged in, they are dependent on large-scale advantages, which mean they have to focus on a limited number of species that can provide the right conditions for their trawl fishery.

The large operation invested significantly in VQS and used their financial power to establish a long fishing season in the Baltic Sea. Based on their own research they began their expansion 2 years before the system was actually in place. Because of the increasing TAC and annual allocations in the Baltic Sea, they have recently bought another vessel (with significant amounts of VQS), which is now catching a portion of their VQS allocation. The large operation is characterized by a two-fold ownership of VQS in the North Sea and in the Baltic Sea, which also constitute their two fishing seasons. Based on their investments in VQS they can run the vessel at full-scale, except for some weeks of holiday in the summer and some days around Christmas. They are dependent on the TAC not only for their annual turnover of kilos but also for the entire value of their company, a risk that is spread over a handful of species. Knowing what they have, the VQS system has enabled the company to build up a brand and to deliver stable landings of high quality. Even though they are mobile, their rhythm and seasons are rather stable. The crew is organized in a rotation system and every person on board receives the same portion of the crew share (which is 34% overall).

General Patterns

The above section presented the different modes of operation in short summaries. In this section, I explicate the common traits and patterns between the operations. The two largest companies clearly saw new opportunities in the system. In fact, they used the new system to reshape and expand their operations. Both of them bought VQS, one enlarged the vessel and the other bought a brand new one. Today both operations are out fishing on long trips almost all year round. This pattern is characterized by expansion and full-scale operation. On the other hand, the small one-man operation adapted and tried to restructure his operations to fit his desired pattern, which was a full year of fishing (120 days) near his home port. In order to readjust to the new conditions, he could change the crew size, vessel, gear type, engage in quota leasing and swapping, direct sale, and change the fishing time. All these factors could be put together to structure a new and perhaps more stable operation in the VQS system. Then there is the tenant, who sold when faced with new investments. He did this to avoid starting a vicious investment spiral, where he would be forced to expand to pay off the first investments. Likewise, the young and new entrant is doing what he can to stay out of debt to the bank. The three are characterized by operations structured to avoid large investments. If the two companies were marked by expansion, then the three others are characterized by a balanced risk-aversion strategy. Seen as a whole, the different behaviors are of course related. They produce the conditions for each other. As the first restructuring took place someone needed to sell so someone else could buy. In the longer run, VQS holders with excess VQS are needed for the leasing system to provide a basis for other operations. Of the ten vessels bought by the large operation described above, six were vessel owners leaving the sector for retirement and two for other reasons. The group of retiring vessel owners has fed significantly into this process of restructuring for both the expansive and nonexpansive operations.

Expansive Restructuring

If we take a closer look at the expansive restructuring, the principle is to provide the basis for the vessel to run at full-scale all year round. Typically, the size and type of the vessel demands that it takes part in fisheries with large catches and high *catchability*. High catchability means that the fish are congregated and easier to catch in great quantities. As fish species migrate, spread, and congregate throughout the season according to factors like weather conditions and spawning seasons, a given species will at certain times have a higher catchability. To optimize the economic performance the expansive operations are mobile and move around according to fishing seasons and shifts in *catchability*. The two operations mentioned above that expanded are both trawlers and, as described, they take part in fisheries that are geographically distributed in the different catch areas. This could be seen as a modern and efficient fishing operation. However, there is a constant risk that it is not possible to obtain the conditions that make this fishing strategy viable. In the following quote, a quota pool manager is talking about one such expansive operation, one which was—at the time of the interview—struggling to make it:

> Then they steamed down to catch cod in the Baltic Sea, and then they were only able to catch 30 boxes per hour, and they needed 100 or 200 boxes before it was viable. Catching 600 kilos per hour was not near enough for them. (Personal conversation, November 2011)

Even though these companies have modern vessels with modern fish finding equipment the unpredictable nature of fishing is still a challenge. The company in the quote above operated with a modern vessel built in 2007, but still the company went bankrupt in 2012. The financial costs are an extra burden on expansive companies, as is the leasing they undertake to fill out the seasons. Central in the expansive full-scale operation is the ability to find and use the unpredictable large-scale advantages provided by nature and technology. In the Danish context, there seems to be a productive combination between the cod trawl fishery in the Baltic Sea and demersal trawling in the North Sea and Skagerrak for plaice and other flatfish. Not only do they complement each other in seasonality (one is in the winter and the other in the summer), they both also often offer the circumstances necessary for profitable large-scale trawling. In addition, the regulation in the Baltic Sea gives any rights holder 160 days of fishing, which means plenty of time to catch the allocated VQS plus what can be leased in. I asked the managing skipper of the large operation described above why companies from far away invested in VQS in the Baltic Sea region:

> They bought it to be able to go to the Baltic Sea and fish, because it was an attractive and easy fishery. Not as high costs as with the North Sea trawling. There is an incredible wear and tear on the gear. Trawl, bottom lines, trawl doors, it is hard gravel bottom you are working on most of the time. Out here [*in the Baltic*] you can run with the same gear, you cannot see it has been used [*because of the clay bottom*]. […] The catch rate [*in the Baltic*] is very high, the few hours we are fishing. In the North Sea we are towing 24 hours per day. Here we are just fishing when it is light, so it's some short days at this time of year [December]. Now we are only active eight hours. We set the trawl when the day starts to brighten, at dawn, and then we can make two drags before it is end of the work day. (Personal conversation, December 2011)

Due to the regulation prior to the VQS system, the Baltic Sea was closed most of the year for large vessels. Therefore, in order to gain access to this "easy" fishery it was necessary to invest in vessels with the specific VQS—the many small gillnetters and trawlers from this region. Therefore in order to obtain a season in the Baltic, a progressive investment strategy was needed, which we heard about in the previous chapter. Today the second largest port, in regard to holding cod VQS rights in the Eastern Baltic, is Hanstholm, located at the opposite corner of Denmark. Stories of Hantsholm skippers walking around in the harbors in the Baltic region with their check books ready, asking people if they wanted to sell—whether exaggerated or not—are a product of this expansive restructuring.

As we know from Chap. 4, the Bornholm vessel owners were not prepared for the race for VQS and were advised to wait until all the rules were in place, while it was the opposite case for Hanstholm (and other) vessel owners, who had done their research and had banks backing them financially. This—a context external to the management system itself—played a part in why it was people from Western Jutland that bought up quota in the Baltic and not the other way around. It should be noted that the large operation described in detail earlier in this chapter, and which indeed has its *home port* in the Baltic Sea, did at that time fish most of the year in the North Sea and was a member of the producer organization in Thyborøn, Western Jutland. Of course for the expansion into the Baltic to take place, this had to be part of the operating plan, using their mobility to travel and take advantage of a prolonged fishing season in the Baltic Sea. But as mentioned above, what characterizes the expansive restructuring is the aim to have the operation running at full-scale all year round, and the expansion into the Baltic was one possible way to achieve this.

Balanced Readjustment

In contrast to the expansive operations described above, the three other operations adapted to the VQS system through a range of rearrangements of their vessels' activities. The one-man operation represents a group of vessel owners in which the gifting of the VQS as valuable collateral enabled a readjustment. But this was not a progressive expansion as described above. Rather this was an investment and restructuring in balance with their new financial situation and fixed allocation of VQS. The new financial weight caused by the VQS made it possible to supplement the initial allocation with a little more in order to secure the base for the production without, importantly, venturing into too risky investments. In the case of the one-man operation described above, readjustment was made partially by investing in some VQS and by changing gear as well as readjusting other factors. In my ethnographic material, other vessel owners have combined the factors in other ways. One, for example, bought a smaller vessel and moved the VQS from the initial vessel to the smaller one. Thereby he matched the vessel size to the VQS and reduced the costs of operation—but this behavior was also prompted by limited financial resources and unwillingness to take risks (Personal conversation, May 2010). The consequence of such a readjustment can be that a colleague is no longer needed in the operation (Personal conversation & 3 May 2010). Likewise, the gear could be adjusted, i.e., using larger mesh size in gillnets to increase the value of the catch (Personal conversation, December 2009). With the VQS allocation in hand, the time spent on sea could be adjusted to the size of the allocation, which would be a very simple readjustment. This could be supplemented with leasing, or perhaps other income sources such as a job on land.

There are many factors that can be manipulated in order to readjust the operation to the conditions in the new management system. In the few examples mentioned above, the factors manipulated are fishing gear, VQS, time on sea, value of catch, and vessel size (costs). What is central to the readjustment pattern is then that the different elements are restructured and put together in a way that can uphold an operation with a given VQS allocation and limited financial dependency. In contrast to the expansive operation, the aim of the readjustment is not to operate at full-scale but to operate at a balanced level—in the case of the small operation described above, this meant a level of "what I can handle" (Personal conversation, December 2011). Likewise, in the case of the young and new entrant—who did not have the advantage of being a first generation beneficiary—there is also the ambition to keep activities at a certain level. They adapt the activities to a level they can and want to handle, and that can support a life as an independent fisher. The aim is to keep the operation at a level that can sustain the livelihoods involved in the operation.

Avoiding large investments is a way to obtain this goal in order for the money to go *in the bank* and not *to the bank* as payments on loans. From the perspective of those in the balanced mode of operation, the expansive companies are engaged in a totally different mode of operation:

> They go where the fish are and then they keep on going and going. That is something we cannot do, we need some more days to catch the fish. We are limited by going out from the shore, but in return we can land fresh fish all the time. (Personal conversation, December 2011)

This underlines the differences in mobility and scale of operation. The large operators can go to other catch areas, but also have to do this. For some, mobility is related to the freedom of being able to stay around the home port. For the young skipper, mobility was a necessary instrument to pick up quota leftovers from his local network—to move around in the North Sea according to quota availability and fishing conditions.

Investment Aversion

At first sight, the investment-aversion pattern is perhaps the most difficult to understand. Investing in the operation to have fishing activities all year round is perhaps risky but recognizable, as we are used to large-scale industrial production in other sectors. Likewise it seems understandable, though slightly more conservative, to readjust to what you have and invest in a little extra, while still avoiding severe debt. But why sell your fishing rights and then lease them back? What explains why the young new entrant is not about to invest in VQS? If he sees a future in the fishing sector, would it not then make sense to build up a quota holding? A part of the answer lies in the fluctuating exchange-value as discussed in the previous chapter; and a part of it is explained by the desire to secure a degree of flexibility and independence.

Flexibility of course means many things and is relative to the mode of production. In this case it is mainly about the financial situation. Buying VQS or investing in a brand new vessel requires significant investments, which in most cases requires bank financing. Building a practice *without* these investments is then crucial to avoid being dependent on the banks. The tenant and the young skipper both do what they can to find niches where leasing and an old vessel can support their livelihoods. What is important for the young skipper is that it is not the bank telling him what to do, but rather that he can make his own decisions. In the interview he joked about the expansive operations as having "HAVE TO" written on their backs—that is they *have to* go out fishing because they *have to* pay the bank (Personal conversation, January 2012). He sees the expansive operations based on external financing as being a contrast to his own operation.

Being dependent on banks is exactly what the investment-aversion pattern seeks to avoid. It is about keeping independence and about freedom to run the operation according to one's own plans and ambitions. In that respect the ultimate flexibility is to be able to stay in the harbor. And in that respect, new investments in quota have to be balanced with the financial obligations they will bring. So far the young skipper has paid in cash and is proud of that. One day maybe he will be ready to invest in VQS—in cash perhaps. More or less on a day-to-day basis he can stop his operation, either for a while until prices (auction and leasing) recover, or he can sell up without a loss. That is, of course, if he can find other options on land.

The expansive operations, on the other hand, gained another type of freedom. With the new market for quota the owners emerged as captains of finance. If the aim is to have the operation running at full-scale, then the financial obligations are a necessary part of this in market-based fisheries management. And with the operation running at full-scale, the freedom *not* to run the operation is simply not relevant. If the owners of these operations emerged as captains of finance, then what about the crew? We know from above that they are not part of the planning and strategic investments in VQS. The tenant is not sure whether the VQS system is good or bad for the crew:

> It is not such a sure thing they make more money, they also get more work. They get more fish through the boat. I do not know if it is good or bad for the crew, I think they were just as well of in the old system. (Personal conversation, December 2011)

Has the breaking down of common and equal access also altered the relationship on board? Have the crew become laborers? With the introduction of market-based fisheries management these expansive companies have taken a step towards full capitalist production. Has market-based fisheries management not only capitalized the resource (as discussed in Chap. 4) but also promoted capitalist organized fisheries with owners, managers, and a crew providing the labor? While a large group sold and retired, those who stayed had to restructure their operations. On one side we have the expansive operations and on the other side a group of self-employed skippers and fishers trying to secure their independence from both the formal labor market and the financial market. One group has embraced the new system and expanded their operations, while the other has tried to balance and limit the necessary financial obligations with their new conditions. In the next chapter, I will discuss the above questions in a more theoretical manner, examining both why market-based fisheries management enables capitalist expansion and why self-employed fishers seem to be in decline.

Chapter 6
Transformation and Modes of Production

Abstract The introduction of private and individual transferable quotas is widely considered to have a negative impact on small- and medium-sized fishing operations. In this chapter, I set out to explore this in a theoretical manner. I discuss the differences in the fishing operations as two contrasting modes of production and examine the ways of life that are enabled by the two modes of production. The central questions are around how market-based fisheries management transforms the principal preconditions for the self-employed fishers; and, in turn, why capitalist organized large-scale fisheries are promoted by this type of fisheries management.

Keywords Modes of production · Capitalist fisheries · Share organization · Individual transferable quotas · Set quantity · Investor capital

Introduction

In the previous chapters I have demonstrated how the VQS as a commodity—with its specific qualities—has had significant implications for the different modes of operation. On one side, the VQS system created the possibility to expand fishing operations by accumulating VQS in order to organize production at full scale. A consequence of this was a greater involvement by the financial sector in order to finance expansion. On the other side, I found a number of operations that sought ways to structure their operation as independent operators. In particular, and in contrast to the expansive operations, financial independence was an important aspect for these operators. For some this was achieved through a limited expansion and readjustment, while for others the leasing system provided access to fishing activities without the need to involve the bank (too much). At the end of the chapter, I argued that these could be understood as two distinct sets of operations and modes of production. In this chapter I expand on this argument in a more theoretical manner, attempting to answer the intriguing question of how market-based fisheries management transforms the preconditions for these two modes of production. To do this, I will have to go more deeply into the concept of mode of production and also introduce the life-mode analysis, which will shed new light on the empirical material presented in the previous chapters and explain how the VQS has suddenly altered the everyday practices of fishers.

The Great Puzzle

Chapter 5 of Thomas Højrup's doctoral thesis is an examination of two modes of production in the ocean-going fishery ("Produktionsmåder i havfiskeriet", Højrup 2002)[1]. The starting point is a puzzle. Why have capitalist organized fisheries struggled so much historically to establish a stable position in the commercial fisheries? How can smaller share-organized fishing units continue to outcompete the large-scale industrial fishery, when in so many other sectors large industrial production is dominant? How can this oddity be explained? This puzzle leads Højrup to a theoretical and empirical examination of the ocean-going fishery. The main question for Højrup is: which relations and connections between different life-modes and their different modes of production can explain this situation?

This puzzle is not just of historical interest. Even in the twentieth century, with all the technological development that came about, the balance did not in many places tip in favor of the large-scale capitalist fishery. This paradox is what Hersoug calls the economic efficiency paradox, namely that (speaking of the Norwegian offshore trawl fleet):

> Trawling was originally considered the most modern fishing technique but it seldom managed to be profitable [...]. On the contrary, the trawl fleet was established with heavy subsidies for construction, was supported with price subsidies for operation and required new subsidies to be scrapped. (Hersoug 2005, pp. 57–58)

As Hersoug implies, subsidies have been and continue to be one of the ways that large-scale fisheries have been established and sustain their position despite their "handicap"—a situation not at all alien to the wider European context. In Denmark even before the introduction of market-based fisheries management, the industrial fishery was one of the most capital-intensive industries across all sectors, only surpassed by the oil and chemical industries (Frost and Løkkegaard 2001). It is relevant and interesting to ask why not only large-scale fisheries often perform badly economically but also, despite this, why they are still promoted by science and management.

A closer look at the two modes of production will illuminate why large-scale fisheries fail to be competitive and indicate why the large scale fisheries are still supported and promoted by management. The free and equal access to the fish resource is intrinsic to this puzzle. Below I will review the analysis offered by Højrup, who compares and discusses large-scale capitalist and share-organized fisheries. This leads to an in-depth analysis of the principles of self-employed share organization, and I will subsequently provide an update based on the experiences from the 10 years of market-based fisheries management that have followed Højrup's analysis.

---

[1] In the following I refer to and examine the analysis conducted in Chap. 5 of "Dannelsens Dialektik" (Højrup 2002, pp. 221–272). The chapter is the most detailed and updated version of material developed over many years (Højrup 1983, 2003). Because of the Danish language I have chosen not to use direct quotes, but I will reference the text when appropriate. The broadest introduction to the state and life-mode analysis in English is "State, Culture and Life-modes" (Højrup 2003).

After that I return to the capitalist fisheries and discuss the implications of market-based fisheries management on this way of organizing fishing activities. Where Højrup tries to explain the persistence of share-based and guild-organized fisheries, one could say that for this chapter the aim is the other way around. Here the aim is to explain why capitalist organized fisheries are promoted and strengthened to such an extent by market-based fisheries management systems like the VQS;—and, consequently, why self-employed fishers lose out.

International or National Agenda

As Højrup finalized his doctoral thesis in 2002, the first transferable regimes for pelagic species were about to be implemented in Denmark. He reasoned that the political motivation for this management change was to mobilize capital for fishing activities in international waters. According to this argument, the political goal was to gain historical rights for the few international species that were not yet under strict quota control. The fishery spokesperson of the conservative party and later Minister of Foreign Affairs, Lene Espersen, argued that the sector needed a capital input to achieve a quick adjustment that could mobilize the fleet to take part in international fisheries:

> There are only a few non-quota species left, and we need the necessary capital and machine power to accumulate historical rights. (Lene Espersen, quoted in Højrup 2002, p. 221; author's translation)

While Højrup emphasizes the political agenda concerned with an international context—and thereby the transformation driven by being a state in an international state system—it is clear now, 10 years later, that something else was also at stake. With the extension of market-based fisheries management to almost every commercial species, we can see that the mobilization of capital and the so-called readjustment of the fleet have also served a larger agenda inside national borders. The sector has been almost entirely transformed and now serves as a platform for expansionist capitalist organized fisheries, better fitted for a capitalist worldview. In this chapter I shall theoretically examine this development. But in order to get there we will have to look into the theoretical and empirical sides of two modes of production in the ocean-going fishery.

Two Modes of Production

In the previous chapter, I discussed two patterns of fishing organizations, which I now examine as two distinct modes of production. One is the capitalist mode, which exemplifies the expansionist, heavily financed industries; and the other is the simple commodity production, which is exemplified by the risk-averse self-employed fishers. In the broadest sense the concept of mode of production is a way to analyze human societies and social relations. The concept is central to a range of academic disciplines that all have their origin in historical materialism and the writings of Marx in particular. It is as such not an integrated theory but more a method to conceptualize how production is structured in principally different ways, and as such it is a method to produce knowledge[2].

Central in the mode of production as an approach is the structure of ownership in regard to the means of production and labor input as well as the appropriation of the end product. A mode of production is a cyclical process that describes one possible way to organize production, including also the reproduction of its own preconditions. Therefore, the concept also points to a range of necessary conditions in the surrounding society or in the social formation as it is termed in this tradition. In the social formation the modes of production are embedded in political, legal, and economic structures as well as at an ideological level. It is also through a social formation that two modes of production can co-exist as a diverse theoretical construction. For example, both capitalist and simple mode of production requires a commodity market where their end products are sold and the necessary material for production and maintenance are acquired. Likewise, a range of political and legal structures is also required for the acceptance and protection of private property as well as people as legal subjects.

Blank Sheet

When working out the theoretical aspects of a mode of production, each mode of production in principle demands a blank sheet and the development of its own conceptual apparatus (Højrup 1983). In practice this means that concepts from, for example, the capitalist mode of production are not necessarily valid for the conceptualization of other modes of production. This of course raises an immediate question of which language and scientific approach should be used in order to achieve such a blank sheet. This is a point worth remembering as we venture into the conceptual construction of the two modes of production relevant for the study of fisheries.

Reading or conceptualizing one mode of production with the concepts from another, results in a sort of centrism, wherein a set of practices is measured from another practice's point of view. This can be difficult to avoid when, for example, applying concepts like petit-bourgeoisie to ethnographic material, where simple commodity producers are seen through the capitalist mode of production. This is also the situation in the fishing sector, where the notion of the small-scale fishery implies a question of degree and size, whereby the small-scale fisher is simply a smaller version of the "proper" fisher.

---

[2] "The problem addressed by the concept of a mode of production is one of producing knowledge rather than classifying data. One looks at societies from the standpoint of a mode of production and from a certain level of abstraction; but this is not to say that additional specificity, that is knowledge at a more historically determinate level, is produced deductively or reductively from general concepts." (Resch 1992, p. 84)

While the capitalist mode of production is well understood and has been theoretically developed, the simple commodity production is much less understood both as a theoretical mode of production as well as in welfare state practices (Højrup 2002, 2003). On the contrary, it is often seen as early capitalism or a traditional way of life about to vanish in modern welfare states (Højrup 2002, 2003). But this is an ideological reading from a capitalist point of view. In this reading it is assumed that the simple commodity producer will either choose to become a wage earner or, if good at their craft, expand to capitalist organized production. But in fact, the simple commodity producer is a much more widespread way of life than is normally considered. Farmers, craftsmen, plumbers, grocers, truck-drivers, butchers, bakers, and many others own and run their own businesses, often in the same complex or building as their homes.

The two modes of production should be seen as two possible, but different, ways to organize fishing activities. In my opinion one of the strengths of life-mode analysis (based on the examination of modes of production) is to point out the existence and persistence of simple commodity production as a fully "modern" way of life that is distinctly different from capitalist organized production (Højrup 1983, 2003). In relation to the ocean-going fishing sector, this means that a definition of the small-scale fishery should not be based on gear and vessel size (length or tonnage), which are the most common ways to define a small-scale fishing fleet. Instead a combination of organizational criteria such as owner operated, guild organized, or the popular but slightly imprecise "community-based" would be more fitting (although of course hopelessly difficult to administer). There is, thus, a need to begin the analysis of simple commodity production from another starting point than the precepts of the capitalist mode of production.

Simple Commodity Mode of Production

As explained above, modes of production are ways to organize production that entail and define conditions for production as well as for their own reproduction. In this regard, the most general conditions for capitalist production are a commodity market and a labor market, from where labor input and raw materials are bought (as well as the end commodities are sold). In contrast to this, simple commodity production has no need of a labor market, as the company owners are themselves the immediate producers. But in line with the capitalist mode of production, the commodity market is also a necessary condition for simple commodity production. It is here that the end product is sold; and also here that the resources for production and reproduction are obtained.

Because of the absence of a conceptual division between a capital owner and a laborer who sells his time, simple commodity production is in its theoretical articulation without a concept for profit and without a concept for labor. These are intrinsic to the capitalist system and to the relation between capitalist and worker. Labor is understood as the producing power acquired on a labor market and as a class of people without ownership of capital that is in principle as much a product of the capitalist system as it is a condition for it. The modern labor market substituted guilds as well as other systems of organizing labor. As the VQS system changed the ownership or access structure of the fishing sector, this is obviously one of the points we will have to return to later in the discussion. The establishment of the VQS system was in all its silence the creation of a "quota-less" class of fishers.

As stated above, for simple commodity producers the external labor input is not necessary, as they are the immediate producers themselves. Likewise, in theory, the production is not directed towards profit, as this would demand a division between the immediate producer and the person appropriating the end product. This does not mean that the simple commodity producer cannot produce an economic surplus, just that this should not be conceptualized as a profit. Similarly, there is in principle, no need to pay wages to anyone, since in theory no one is hired. This is also why the life-mode analysis only finds one life-mode based on the simple commodity mode of production—at least at this level of specification. Simple commodity production is often described as organic, referring to the nondivision between (in capitalist conceptualization) labor, the company owner, the manager, and the individuals appropriating the end product. Where we have wage earners, managers, and investors as separate life-modes in the construction of the capitalist mode of production, there is only one life-mode in the simple commodity production, namely the self-employed.

Specific to this life-mode is that the self-employed workers do not get paid in exchange for their labor but rather receive something very different as a result of their effort. The simple commodity producer produces his or her own product, and therefore the conditions for his or her own existence. Hence, there is a circular relation between means and ends, which is both to sustain the unit of production and to be able to produce. It was this circular relation that was present in three of the modes of operations described in the previous chapter. One of the primary incentives of their operations was to stay independent as operators and not necessarily to expand and circulate as much capital as possible.

The Commodity Market

The starting point for the construction of the simple commodity producer is the commodity market and the production for this market. This is because the most basic relationship in this mode of production is between production of commodities and the sale of these on the commodity market. Therefore, in order to continue to exist as a producer, the income from the sale has to be at least the same as the costs of production and reproduction of the unit. According to Højrup (2002, p. 238), this relation between income function (commodity market) and cost function (production) is the main theoretical framework for understanding the simple commodity producer.

The income function depends on quantity and market prices—in this case the fluctuating prices on a globalized market for fish products. A fisher can increase the outcome of the income function by increasing the quantity of fish caught, not by setting the price on his or her own. The cost function is the cost of operation; it consists of overheads, which are the basic costs, and the quantity related or variable costs. Examples of the former are vessel maintenance, loan payments, and fishing gear, while examples of the latter are oil, ice, as well as the food consumed during fishing trips. The central entrance point for the construction of the simple commodity mode of production is then the mutual and reciprocal modification of the cost function to fit the conditions of the income function. This is the most general aspect of the simple commodity producer.

This understanding then leads Højrup to sum up the contrast between capitalist and simple commodity production (2002, p. 239). In this summary, the capitalist mode of production consists of the distinction between capital, labor, and means of production; the simple commodity production consists of the distinction between overheads (basic costs) and operating costs. Both produce for the commodity market and are dependent on this for income and the purchase of goods for production (and hence they can co-exist), though they differ in their mode of production. When the simple commodity production is employed as a fishing practice it is most often in the shape of share fishing. Thus, the overheads mentioned above contain both the fixed costs of the vessel as well as the proper income for the crew. This is expressed in the share system, with a boat share and crew shares. The income from the sale is shared after the variable costs—oil, food, and ice—have been deducted.

Fixed and Variable Costs

Fixed costs are related to the cost of the equipment that enables the production—the overheads. In fishing these are, most importantly, vessel and gear; but they also include an array of other costs, such as insurance, memberships of producer organizations and harbor fees. These have to be paid even if the vessel stays in harbor. These reflect the level of operation, i.e., the magnitude of the fishing activities. The size of the vessel, the number of gillnets, or the size of the trawl set the level of the operation, the average quantity of fish that the operation will have to catch in order to balance the cost, and income functions.

The moment the unit goes to the sea, variable costs will begin to accumulate. The longer the trip the more oil, ice, and food will be needed, which will result in rising costs. The rise in costs is hopefully balanced by a greater catch, which brings us back to the central definition of the simple commodity producer: the balance between the income function and the cost function. The main point is that the reproduction of the means of production, understood as an organic whole, is the goal of the catching unit as well as the internal requirement for its existence. For the self-employed fisher the goal is to: "maintain flexibility in order to stay independent" (Højrup 2002, p. 241).

If we bring VQS into this analysis, we can see that when a simple commodity producer leases VQS, it is a variable cost. If VQS is bought, repayments on loans become a fixed cost, to be paid even if the boat does not leave the harbor. This goes against the primary incentive of flexible and independent reproduction of the catching unit. Thus, the life-mode analysis recognizes why many fishers prefer to pay more to lease rather than buy quota, a behavior that is not "rational" by the standards of mainstream economic theory. This is because by applying the rules of mainstream economic theory to simple commodity producers, we are applying the standards of one mode of production (the capitalist mode) to a completely different one (the simple commodity production mode), with an entirely distinct internal logic based on flexibility and independence. Leasing is a way to maintain flexibility and independence, though it might be more rational in a long-term, large-scale but intrinsically capitalistic perspective to invest in the required quota.

What the self-employed fisher can do in order to stay independent is to manipulate the elements in the cost and income side of the equation. Højrup develops four basic ways of reasoning or strategies to explain the daily practice and operation of a simple commodity producer (not just fishers), each based on one of the manipulative elements. These four elements are price manipulation, manipulation of quantity related costs, manipulation of fixed costs, and manipulation of the size of the catch (quantity). These should not be seen as mutually exclusive but rather as different manipulative aspects, strategies, and ways to organize production.

Price Manipulation

For a fishing unit it is, in principle, not possible to manipulate the price. Price manipulation is a manipulative property reserved for "traders." The trader must set the prices at a level that will cover the overheads and variable costs in the long run. Since, according to Højrup, the fisher cannot manipulate the price, this is not part of a possible practice for a fishing unit. Perhaps this is a point where the VQS system has changed the situation slightly. In the material described in the previous chapters, there are a few strategies that indirectly influence the price of fish. The large company worked hard to build a brand and made sure that the fish were treated gently during production. Because of this they received a higher price than others landing the same fish species. Here it is important to note that they basically improved the quality of the product, and therefore landed a different product (premium quality fish). Their price per kilo of fish was thus not higher, while the price per kilo VQS was. The same can be said of the young man who switched to another port and auction with higher prices. Both of these behaviors were still dependent on the prices offered by the buyers. The VQS system may have enabled these shifts in quality orientation though.

Selling the fish directly to the consumer probably qualifies as the best way to manipulate price, as in the case of the one-man operation that sold to tourists and locals. The amounts sold like this are of course low, and it could be argued that the product sold is more than the fish; it also includes an added value of an immaterial experience for the consumer. Thus the product is different, not the price. In some way the higher price technically derives from shortcutting links in the distribution chain. Perhaps direct sale could best be conceptualized as the producer taking on a new role as a trader—where websites, text messaging services, and so on complement his production unit, now as a trader.

The VQS system has created a new heterogenic relation between the VQS value and the landed value. It is, in other words, possible for some to get more money out of one kilo of VQS than others, as indicated in the following quote, where the skipper of the large company explains the effects of their branding strategy:

> We can sell together with the others and get 20% more than they get per kilo, even though they were all caught in the same place. It is a brand we have built up over several years. (Personal conversation, December 2011)

In management systems not based on market mechanisms there is of course also good and bad quality; but the introduction of the fish as a VQS commodity has also established a new relationship between the value of the VQS and the landed value of fish. This is not only due to the handling of fish on board but also choices about when to go fishing. With individual shares (not necessarily transferable) it is possible to distribute the catch in closer coordination with the market prices.

Manipulation of Costs

In the second strategy manipulation is directed towards the quantity related costs. This is what in general can be termed a capital intensive strategy. To lower operating costs it is necessary to invest in the operating equipment. In the case of fisheries this would be to invest in a larger vessel or in more nets in order to increase the volume of the catch. When the operating costs are lowered by investments, the quantity of the catch has to be increased in order to pay off these investments (and the consequently higher fixed costs). This is sometimes termed volume-fishery.

If the situation is the other way around and the overheads are manipulated, the overheads (and therefore the fixed costs) are kept low in order to keep operating even with small catches. This third rationale is also a typical strategy in fishing, where a small unit with low fixed costs can sustain a practice even with low quantities. Typically, this would also be visible as a change related to a reduction of financial pressure. When the boat is paid off the intensity of fishing—the number of days at sea and the duration of the fishing trips—can be kept at a lower level, without necessarily resulting in a lower personal income. The knowledge of an experienced fisher regarding seasonal changes, gear, and the best fishing spots also contributes to "slowing down" as a viable strategy.

Quantity Manipulation

The fourth way of reasoning is for a fisher to increase the quantity of fish caught in order to make the income function cover the cost function. We know that the most basic feature of a fishing unit is its need to keep the income function above a certain level in order to cover costs. Without price manipulation, quantity manipulation is therefore the most general aspect of a fishing unit—or was. When Højrup published his doctoral thesis in 2002, fishers had privileged and free access to fish resources. For the active, licensed and registered fishers, the main ingredient in the production of "fish" was cost free. This is perhaps one of the most crucial changes under the VQS system—as quantity can no longer be manipulated cost free. The leasing market is the only way to temporarily manipulate quantities by accessing or leasing out quantities of quota.

The Catch Unit Further Specified

If we return to the analysis offered by Højrup, central to the practice of the catching unit is that it has to land a quantity of fish in kilos that results in a revenue that is larger than the fixed costs (Højrup 2002, p. 255). In other words, after the variable costs are paid the revenue from the catch has to cover the fixed costs over the course of the year. Therefore the central reason for a fishing unit to be engaged is to fish long enough to cover the fixed costs. This can be manipulated both in relation to the duration of each fishing trip as well as the number of them. To understand this, it is important to remember that in a share-organized fishery the labor is not a variable cost; it is not bought through a labor market. So when fish prices are low the unit can be engaged in longer trips and in more of them in order to increase the quantity, without increasing the labor costs. It was in this way that the manipulation of quantity constituted the most central element in a simple-commodity fishing unit in a regime of free and equal access.

We can deduce from this that in the VQS system the operator either has to manipulate the elements in the cost function in order to adjust to a set quantity determined by the VQS allocation, or manipulate quantity through investments in VQS or through leasing. Both ways of manipulating quantity will bring new costs.

Maximum Production

Before VQS, when the boundary for normal production (as determined by factors like vessel size, weather and physical health) had been reached, the unit either had to invest in operating equipment to be able to increase the catch or try to lower the variable costs. The former would be investments in more gear or a larger vessel. The latter could be done by increasing the combined knowledge of the personnel on board through searching for new fishing grounds where unused aggregations of fish lower the operating costs. The wreck fishers of Hvide Sande were an example of such experimentation, which involved development of both technological and ecological knowledge (see Chap. 1).

Normally, neither innovations nor new fishing grounds provide exclusive knowledge for very long. So in the longer run this process creates both a common pool of knowledge for the whole fleet, and at the same time instilling an internal drive for further innovation and experimentation in the sector. In time, any competitive advances a fisher makes by lowering the costs of production will be assimilated by the rest of the fleet. It is questionable how this process is changed by the introduction of market-based fisheries management. Market competition and the advantages of lower-cost production still play a role, but the methods of attaining these have changed, as a set quantity of catch is now the point of departure for the fishing unit. I will return to this later.

Switching Fishery

Another option specific to fishing practices is to switch to another fishery where the quantity and price relation is different. Switching like this requires that the operational equipment, like gear, vessel, and knowledge can be employed directly in this other fishery or at least with a minimum of new investments. There are, according to Højrup, two variations of this reorganization. One is to switch to another species that is less abundant but attracts better prices, while the other is to deliver a higher quality or better processed product. Still, the practice revolves around the basic principle of keeping the quantity of the catch high enough to cover the overhead costs. Obviously, switching to another species is difficult in a market-based regime, where shifting fishery would require either investments in VQS or venturing into the leasing market. As I have documented in the previous chapters, leasing is widespread and fully integrated into fishing operations. This indicates that switching is still a component of basic flexibility, but now with the extra cost of lease rent—and with the varying opportunities coming from the dynamic leasing market.

A new problematic aspect of market-based fisheries management is that costs from earlier investments in VQS still have to be paid off, even if allocations are much smaller than in previous years or not caught at all. This has likely been the case for those who invested, for example, in Kattegat cod in the early years of the system and then experienced a decreasing TAC in subsequent years. The flexibility of the simple commodity unit to switch between species is thus changed under VQS, as it necessarily involves leasing or investing in VQS. On the other hand, the VQS system better allows the option of improving the product and landing fish of a higher value. There is no closing of fortnightly rations and a whole year to catch fish at the highest price, its best size, or to make deals with local restaurants. This introduces a new kind of flexibility.

Normal Production

According to Højrup, the simple commodity producer (prior to the VQS system) would seek to find a normal level of production to ensure the continued reproduction of the production unit. Based on expected prices, quantities, and the estimated fixed costs, the unit seeks to organize the vessel and production to fit a normal level of production. But it will also seek to consolidate and invest in means of operation so the unit can be operated cheaper or with a higher catch rate. Thus, according to Højrup the units and the fleets as a whole increase their capacity stepwise, by investing in and improving the vessel as well as by building new vessels (Højrup 2002, p. 251). The simple commodity producers seek to improve their resilience to ecological fluctuations and the fluctuations of prices on the market. This is a process based on free and equal access, and it is partly also an answer to the technological developments and constant competition in the world market described above. The implications of the VQS system are to obstruct this process and force the units to adapt their costs to the given situation of quantities and prices.

Vessel Circulation

These constant adaptations and switching between gear, target species, and vessels created a dynamic market for vessels. Through a vessel's life it shifts owners, changes shape and names, and takes part in fisheries of very different kinds and from changing ports. The different sizes, riggings, and capacities, as well as the cost differences between vessels in good and bad conditions, support the plurality of ways to organize production in the fishing sector (Højrup 2002, pp. 249–250). Thus, the average age of a fleet does not say much about its utility nor its condition. An older wooden trawler can serve very well as a good gillnetter in a place distant to where it was built before, perhaps, ending its days in another country. However, the VQS system has altered this situation by changing the dynamics of the vessel market. Vessels with VQS attached have increased significantly in price, while the market is to some degree flooded with vessels without any fishing rights. In that way the VQS as a right has become a property of the vessel and influences its lifecycle as an immaterial component. In regard to the vessels with VQS, the value of the material vessel often represents a minor part of the price. The VQS and other rights take up an average of 80 % of the value of these ships, according to shipbrokers I have asked (Personal conversation, May 2010). It is not uncommon that vessels are now traded because of their attached VQS or simply just used to move VQS from one owner to another, as we saw in Chap. 2. On the other side, the reduction in operators caused by the VQS system has created an excess of vessels, which opens up new possibilities to buy a vessel at low cost and build a fishing practice on leasing.

Constant Adaptations and the Share System

Above I have sketched out two main features of the simple commodity fishing unit: the manipulation of quantity and investments or reorganization in the fishing operation in order to manipulate the fixed and variable costs. I have pointed out some aspects that were influenced by the introduction of the VQS system and areas where life-mode analysis allows us a greater understanding of the logics of simple commodity production than contemporaneous economic theory. In Højrup's account, human organization is also important to the flexibility and adaptive nature of the simple commodity unit. Share organization is one way to achieve this. In order to make adaptations and switch between fisheries, the whole of production, including the people, have to be organized in a way that enables these changes. This means that the knowledge, vessel, crew, fishing gear, work relations with other units, electronic navigation and fish-finding equipment, as well as financial matters, have to be organized in a way that enables constant and dynamic changes in catch quantity, species, fishing destination, vessel length, and number of fishing trips.

The share system is, in other words, one of the possible ways to organize a flexible and adaptive fishing unit. The basic principle is that the totality of means of production—which in this mode of production includes vessel, gear, and people—receives a share in the revenue according to the costs of reproduction. Based on the expected annual turnover the total income can be estimated, and it can be worked out how large a share of this will be needed to cover maintenance of vessel and gear. The concrete size of the "boat share" depends on the size of the investments (in vessel and gear) and the maintenance costs, typically ranging from 25 to 75%. The remaining part, the crew share, is then divided in equal shares to the people on board (with the exception of smaller shares for untrained new crew members). With the risk of loss shared, and since there are no set wages to be paid at the end of every month, the share system is robust and resilient to both market and ecological changes.

Ideological Relations

Share organization should also be understood as an ideological way of life; that is, as a mode of production characterized by a shared ideology, rather than an economic and contract-based way of life. The reason and meaning of their cooperation is not economic in the sense that one pays the other, or political in the sense that they are legally forced to work together. This requires a shared "team spirit" in decisions and actions—a shared ideology and mutual cultural understanding, which in some areas transgresses the boundaries of a working relationship into private life and being part of a community. Since the crew is made up of a number of local fishers, sometimes changing in size according to season or ambitions, ideological relations are rather central in structuring the shared fishing unit:

> Whereas the wage system may be said to provide an organisational structure for a capitalist enterprise, the core of simple commodity production is not organised around economic relations but rather on the basis of social relations in the form of family ties, cooperative relations between partners and other ideologically based associations which bind producers together into cohesive production units. (Monrad Hansen 2012)

The relations between fishers are marked by the fact that they are potential partners or have been partners in the past, if they are not cousins, neighbors, or related in other ways. In addition, the fishing units are part of a competitive but cooperating group of fishing units within a community: they compete for fish but also cooperate through information sharing and planning of fishing activities, i.e., harbor investments (Andersen and Wadel 1972; Löfgren 1977). The important knowledge about fishing grounds, gear, seasonal differences, fish migrations, and so on are passed on from partner to partner, not from manager to crew. This means that a certain shared ideology is negotiated and developed for the group of fishers, families, and their community. The shared ideology comprises a multitude of ideas about what a "real" fisher is, what is accepted behavior (on land and sea), and what the goals of the community are in relation to individual benefits and ambitions. The special mode of organization, with its ideological side, gives rise to different "ways of life" and "habits of the heart." While the share fisher puts pride in "being able to independently plan his own life together with his partners on board" (Andresen and Højrup 2008, p. 31), the wage-worker is fully dependent on the employer. Empirically there are arguably different degrees of ideological unity among share fishers. In some communities operations more or less resemble one big family, while other units have relationships closer to that of boss and hired workers.

Coming to a conceptual understanding of the ideology and way of life in the share-based mode of organization as well as the cultural meaning of share organization—often expressed in words like "community feeling," "social cohesion," or the concept of "social capital"—is central for our inquiry into market-based fisheries management. Seen from an ethnological point of view, it is exactly this aspect that political decision-makers and economic analysts failed to see and cater for. The community aspect is not pure nostalgia or rural romance; it is an important and significant part of a contemporaneous mode of production. This is the core of the disciplinary gap. What the ethnological researcher uncovers through qualitative fieldwork is deep transformations caused by market-based fisheries management.

The VQS complicates the relationships between the people involved because it suddenly hands ownership of the resource and raw material of production to the vessel owners and creates a class of nonowners. Vessel owners can raise the "boat share" unilaterally, simply because they can. The quantities caught are no longer a shared decision and responsibility. The ownership of VQS could perhaps be compared with the ownership of the vessel, which is often also owned by one person. However, not only does the VQS represent a much higher value and have no costs of reproduction for the owner, but the crew have also had their principal access confiscated.

In the above section I have examined Højrup's predominantly theoretical account of the simple commodity producer as a fishing practice. By contrasting this with findings from my case studies, I have been able to point out where the VQS system has had principal implications for the producers. In particular in relation to the quantities and species a fishing unit can target, the new regulation has brought significant changes. In the following section, I attempt to summarize the aspects of simple commodity production that the VQS system has challenged and changed. In addition, I outline and discuss the new characteristics of the simple commodity production in market-based fisheries management.

Discussion: Principal Changes with the VQS System

I have examined the simple commodity mode of production and its manipulative elements as presented by Højrup (2002). An empirical and theoretical precondition of Højrup's account was free and equal access to the fish resource, an access structure which is no longer present in the Danish commercial fishery. In Højrup's account the catching unit was further specified as, essentially, a quantity-manipulating unit that had to adapt to shifting conditions in both nature and market through changes in quantity and target species. Understood through this lens, the practice was concerned with keeping the quantity at a level that would ensure its reproduction.

Vessel, crew, and gear had to be organized in a way so a certain quantity of catch could be reached. Therein lay flexible shifting between fish species and fishing grounds as well as vessel size, gear, and the number of people on board. The share organization was then proposed by Højrup as one possible way to organize such an adaptive and flexible unit. In the share organization every person and material input is paid according to its costs of reproduction, and the producers were tied together by ideology—understood as a social relation—characterized by a shared cultural meaning rather than a legal or economic relationship as, for example, the wage structure. In short, that was the account based on a free and equal access structure. If we re-examine the simple commodity production, now under a market-based multispecies management regime, the most obvious change is that the manipulation of quantity and shifts between target species is no longer equal and free. Historically this has been the foundation of simple commodity fishing units.

Set Quantity as an Entry Point

For the simple commodity producer this means that the flexibility to freely switch between fisheries and to increase or decrease catches is no longer there. Instead the quantity is now given by the ownership of VQS and the politically determined TAC. Therefore, in order to change the quantity, the simple commodity producer is forced to either buy or lease VQS. The first would in most cases have to be financed through bank loans, while the latter demands cash enough to pay for the leasing up front. In this respect the VQS system brings about greater involvement in the financial system both in the long and short-term practice.

For those units that are not engaged in the leasing market, the consequence is that they have to be understood in the short term not as a manipulator of quantity but in relation to a set quantity. As the quantity is not manipulative, the unit therefore must adapt its cost function to the expected high and low levels of the income function. The fishing unit has to adapt in theory to a situation with both low TAC and low market prices. In consequence, the operation has to be downsized in order for the fixed costs to be covered by the income from the fixed quantity of VQS.

For those units that use the leasing system to increase their activities, the operation must be arranged so the extra variable cost from leasing still leaves a crew share large enough for the fishing practice to be economically attractive. Two of my cases used rather old vessels, which they had acquired without the need for external financing, in turn keeping the boat share down. Others have chosen to invest in VQS to increase their allocations; but they still have to adapt to a set quantity, which might leave them more operational freedom but also higher fixed costs due to the financial costs. The freedom they gain with higher quantities is evaluated against loan obligations.

Crew and Share Organization

What happens to the share organization when suddenly enrolled in a market-based fisheries system? In the share system each part of the production received a share of the income generated that equaled its cost of reproduction. In theory, the people involved shared the potential benefits and risks in going to sea. In relation to the vessel owners that choose to acquire further VQS, there will typically be an increase in the fixed costs, as the loan for this investment has to be paid off. This can be charged to the crew as a lease fee or included in the boat share (Personal conversation, January 2012). In the first case the VQS owner is extracting a rent that marks the new unequal relationship between owner and crew. In the latter case the investments should then be considered a part of the reproduction costs for the vessel, where financial costs are financed through the "boat" share. The boat share served to sustain the boat's value and use in order to remain flexible, switch to another fishery or sell the vessel to adapt to changes in another vessel.

There are, however, aspects to the quality of the VQS unlike those of the vessel. While the VQS enables the unit to take part in fishing activities (like the vessel), the VQS has no costs of reproduction. Even though the VQS represents an amount of fish out in the ocean, as an investment it resembles a financial asset more than a physical part of the production. It has been capitalized, and consequently it is a financial element. The VQS as an immaterial value has been implemented into the practice. Compared to the vessel, the VQS can store value much better that can later be cashed in through a sale. In that case the individual ownership of the VQS is important. In the case of a sale, the income (or loss) from the VQS is individual but has been produced by a group. There is, in other words, an appropriation by the VQS owner of the value created collectively; and this, in theory, is a break from the principles of share organization. The situation now resembles the theoretical relation between the productive capitalist and the laborer.

In other words, to use a cliché, all share fishers are equal but some are more equal than others. Thus, in its daily organization the changes to the share system are rather small; but in the case of a sale, the accumulated value in the VQS is privatized to the individual vessel owners. The crew might experience a small increase in the boat share, but with a larger production their income remain the same or larger (Personal conversation, January 2012). Their work is slowly transformed and layered in the exchange-value through the annual use of the VQS. On the other hand, if the unit does not reach its normal production it is the vessel owner who is left with the demands for bank payments. In that way the VQS system forms and demands an individual subject behind the VQS quota.

Withering Away

In many ways the share system revolved around a "we," which continually adapted its organization as a unit and as a whole to the given conditions. The VQS system broke apart that we, or at least made the first cracks. The question is how it will look in the long run. The ability to adapt to constant fluctuations and to reorganize the units in order to meet increasing demands has been severely limited. The simple commodity producers can choose between investing, leasing or downsizing their operations. Downsizing means going from larger to smaller vessels—from three to two people, or from two to one-person operations—from full-time to part-time, and so on. At the same time, the dependency on leasing will increase, as long as the leasing market can keep on sustaining the many gaps in production.

Alternatives include, of course, increasing the income by direct sale and similar product-based alternatives. Along with the possibility to better plan the fishery, these are positive elements of the VQS system. Regardless, the result is a mode of production that is locked in its ability to compete, unless competition is taken up on the financial market. With the increasing involvement of the financial markets comes also greater individualization, as the financial systems to a great extent are oriented towards the individual agency. Those who do invest have to compete on the basis of VQS prices set on the market in competition with larger operators; and this can result in increasing financial obligations, perhaps worsening their overall conditions. With the current prices, the growth of the simple commodity fleet (seen as a whole) seems to be suspended. And seen in a long-term perspective, individual actions to uphold economic viability through adaptations and downsizing could result in a withering away of the simple commodity producers in the fishing sector.

Summed up, in the VQS system with fixed quantities of quota, the producer must adapt his or her costs to the fluctuating annual quantities of TAC, leasing costs and market prices. In the worst case scenario the producer has to be able to survive a year of decreasing allocation and low market prices, and the operation must be arranged to cater to that situation. This is the consequence of being locked in regard to both quantity and species. The leasing system can provide some of the flexibility for the operators, as they can lease in extra amounts as well as amounts of other species, but the fixed costs as well as the variable costs have to be low enough for this to make economic sense. Selling the quota to pay off the house and the car might make sense in this understanding, if the continued fishing activities could then be carried out more independently. In this example, leasing is of course a new cost; but the fixed costs—including those of the household—are now much lower. Thus there is a built-in process in which the simple commodity producers that have adapted to the new system are also locked in their size, and contrary to earlier systems they cannot organically grow and adapt the size of their production. When set for sale, they are accumulated by the captains of finance from large and financially stronger companies. That explains why self-employed catching units decline in a market-based fisheries management system. But what enables capitalist fisheries and the captains of finance to prosper?

Capitalist Fisheries

The central relationship in the capitalist mode of production is between the actual producers and the owners of the means of production and the produced commodity. In capitalism the capitalist owns both the means of production and the end product. Labor is bought by the capitalist through the labor market and is as such under the control and management of the capitalist. The worker has to sell his or her labor and time in order to receive a wage to live from. Since the worker is in principle without ownership of the means of production, the labor market is a necessary economic condition for the worker as well as the capitalist. At this level of specification, life-mode *analysis* attributes two life-mode concepts to the capitalist mode of production. These are the productive capitalist and the worker, each with its own relations to production and to each other.

Another central element in the capitalist mode of production is surplus. As the capitalist can produce more value from the labor than is reflected in its exchange-value, a surplus is created that belongs to the capitalist. This is a unique capacity of labor and of the labor market, and it allows the labor input to have a higher use-value than exchange-value (the costs of its reproduction), which in turn can yield a surplus or profit appropriated by the capitalist. Where the worker sells his labor time for the capitalist to use, control and put to work, the capitalist has a double role as both the owner and the manager of the means of production. In advanced productions this double role is dissociated into two different life-modes: the investor and an employed manager.

The manager is the person in charge of the production who controls the labor, while the investor is the absent owner. Crucial to an understanding of the dynamics of the capitalist mode of production is the internal competition between capitalists. If it is not possible to organize production in a competitive way and yield a surplus, the production will be closed and the investments drawn out and moved to other productions with a higher surplus. This competition puts pressure on how the production is organized and optimized, and it introduces a possible contradictory relation between the investor and the manager. The investor can force a sale or closure by moving his capital away from production. The investments are in theory constantly compared to the average rate of profit produced by other capitalists and the international system.

With the ability of simple commodity producers to increase the quantity of the catch, and as the fleet has grown and adapted to the given conditions, it has not been possible to uphold a rate of profit high enough to allow capitalist producers to meaningfully obtain full control of the fishery. In a situation with free and equal access, and with enough potential fishers with ambitions of being self-employed, it is not possible to create a situation with significant surplus (Højrup 2002). The investments will mean further capacity or new units enter until the surplus is gone. This process is important in understanding the economic functioning of the capitalist fisheries under equal and free access, as it is difficult to uphold a competitive production compared to the average rate of profit.

The main theoretical components of the capitalist mode of production are therefore constant capital (raw materials and means of production), variable capital (labor) and surplus, while the competition on rate of profit is a central organizing principle. Each time a commodity is sold on the market the capital created flows back to these three. The raw materials and the worker are paid for, and the investor gets an economic yield from his or her investments. The capitalist mode of production secures a constant circulation of capital and commodities and is in addition to the commodity market dependent on both the financial market (for credit) and the labor market (for labor and mass consumption).

Central to the labor relation is thus the appropriation of the end product and the circulation and accumulation of this as capital. Based on its use in production, this capital can be split into circulating capital and fixed capital. Fixed capital is the operational equipment that is used in several cycles of production. The circulating capital are raw materials and other operating expenses. To reproduce the production unit, the aim is to maximize the amount of capital circulated in production, while the fixed capital defines the production span and uniqueness of the product (Højrup 2002, p. 258). Herein is an incentive for full-scale production, as the circulation of capital is the principal method of yielding an interest on investments. Thus, the more fish that can be caught and produced by a vessel per year, the more capital can be made profitable—or rather, the capital is made more profitable. When the capitalist fishing company invests in fixed capital, i.e. in a larger vessel or more gear, the aim is to engage as much circulating capital as possible and to do this at a high rate of profit.

With free and equal access to a limited resource, the main raw material and circulating capital is subject to competition and scarcity. This has negative effects on the market prices when supply is too high; it can also result in higher costs of production as the fish is harder to catch. However, through market-based fisheries management the main resource can be monopolized and operated according to a rational production plan. As the capitalist producers are subject to constant competition, though, accumulated capital has to be reinvested in order to stay competitive. This is what David Harvey calls "the capital surplus absorption problem" (Harvey 2011, p. 26). If the capitalist does not invest and improve the production someone else will, and the production will in time cease to be profitable. Thus the capitalist mode of production has several inherent dynamics that push for constant expansion and growth, which is also evident when looking at empirical developments of capitalist industries.

Looking back at the management of the Danish commercial fishery, the limitations on new capacity and growth implemented in the 1980s and 1990s were a severe obstacle for capitalist growth in the fishing fleet. Without these restrictions the technology-intensive pelagic fishery was, for example, a type of fishery that could be almost monopolized by large-scale capital. With the large trawl and purse-seining vessel employed in pelagic fishing, this monopolization occurred through high entrance costs and not via the exclusive ownership of the resource. On the other hand, the introduction of a market for fishing rights suddenly re-introduced the possibility of capitalist growth in the sector and created the potential to absorb surplus capital through the new VQS commodity. The operations could subsequently grow both in size and in quota holdings, which secured circulating capital for production and new large-scale advantages. To illustrate this, we can look at a concrete example.

Isafold

One of the largest fishing vessels in Denmark is the 76 m long HG 333 Isafold. A Danish newspaper estimated in 2011 that the company, including its fishing rights, had a value over 1,000,000,000 DKR (approximately 175,485,000 US Dollars). Isafold is a pelagic trawler and purse seiner that left the shipyard in May 2006. Prior to this, the company that owns Isafold had been running a total of four vessels in the pelagic fisheries in and around Denmark. When market-based fisheries management was introduced for herring in 2003, the company's activities were soon consolidated from the four vessels into the newly constructed Isafold. The company that today owns Isafold began its fishing activities back in 1975 as a project initiated by the shipbroker company "Niels Jensen & Co" (established in 1972). The reasoning behind this new fishing enterprise was based on declining landings of herring in Hirtshals. The decline was due to the expansion of Exclusive Economic Zones that had expelled Icelandic purse seiners from the North Sea. This led to a decline in the landings of herring, and the local fishmeal and oil processing facilities in Hirtshals thus had a shortage of raw materials for their production. In other words, the company saw an opportunity for a "modern herring vessel" to fill the gap left by the Icelandic purse seiners.

At that time, purse seines were relatively new to the Danish context, and together with an Icelandic skipper and some local investors the first *Isafold* (the precursor to the current one) was contracted. The main fisheries back then were herring and mackerel, but Isafold also took part in the pioneering blue whiting fishery west of Ireland. Isafold was successful and the company soon ordered another larger vessel, "Geysir," which was, however, delivered shortly after the herring fishery was completely closed in the North Sea in 1979. In 1985 one of the passive investors wanted to leave, and as part of the exit deal the company took over another vessel, HG 224 Fabian. However the Fabian, a 35-m vessel built in 1955, was not a good investment, as it was too small to be profitable; as a consequence it was hard to retain crew. The company also bought shares in HG 262 Lene Polaris (36-m), the economic performance of which was also unsatisfying.

Limits to Growth

With the introduction of the Common Fisheries Policy in 1983 restrictions were introduced on new capacity, and further licensing and quotas were installed to regulate the fishery on all main stocks. The Isafold company was therefore restricted in its expansion, as the capacity regulations limited the company in its growth and locked the vessels in size. As a consequence of the new regulations, even Isafold was quickly too small to take part in the competitive blue whiting fishery. This was when the blue whiting fishery and others were finally closed and the quota shares subsequently distributed based on historical catches. Consequently, Denmark lost access to the blue whiting fishery because of lower historical catches in those years. Here we see the problem Lene Espersen referred to at the beginning of this chapter that was the motivation for management changes, according to Højrup (2002). The regulation limited the Danish companies in their growth, and consequently Denmark was losing ground in relation to competing countries. Thus, the company was locked in its 1980s size until the introduction of ITQs in the pelagic fisheries in 2003.

Based on this reform the company could reorganize its activities. In a very short time the remaining shares of Lene Polaris were bought, and all quota shares were immediately transferred from the two smaller vessels to the larger Geysir and Isafold. Finally, a few years after the introduction of ITQs, the new Isafold was built and all "quota" and fishing activities were consolidated in the 76-m long vessel. According to one of the partial owners of the new Isafold, its size enables it to be at sea in almost any weather condition without risking the safety of the crew (Tarbensen 2012). But even with the consolidation of four vessels to one, there was potential for further expansion. In 2008, Isafold used only 130 days to catch the total quota originally attributed to four vessels. Therefore the company saw an opportunity to expand, and in 2010 they bought further quota to supplement their fishing activities.

Organization on Board

The fishing activities of the Isafold are organized around a ten-man crew consisting of:

- A skipper
- Two officers (one of them second skipper)
- Three engineers

- A chef
- Three fishers

The crew rotates on a one-to-one basis so there are a total of 20 people involved in production. Measured in tonnage, Isafold is almost three times larger than the older Isafold, though it only employs one more person than the former vessel. By consolidating the business into the new Isafold the number of people employed has thus been reduced significantly, and the crew from at least three other vessels has more or less been made redundant. The size and investments in technology on Isafold allow for enormous catches and annual turnovers. In addition, the new vessel has 14 crew cabins—each with toilet, shower, TV, and access to the Internet—as well as the ship's galley, mess and an exercise room. Another feature is that, with the new consolidated unit, "energy" has been freed up for education and courses for the crew, and the company collaborates with representatives from the working environment board. The administrative burden surrounding the vessel is significant, especially the consuming task of readying licenses by the beginning of the year. To aid in the administrative tasks, the company has a management team on shore.

The immense value of the company and its expansions require large investments. The group of owners now consist of the initial founder (60/200 ownership), the skipper (66/200 ownership), the chief officer (8/200 ownership), and the shore-based manager (66/200 ownership). The latter two are part of a generational shift, in which a previous investor and the founder sold shares to the young chief officer and the shore-based manager. Compared to most of the other examples I have discussed, Isafold is a very large operation. The scale is now so large that each work-function on board has been made into a separate position, covered by one to three crewmembers. Cooking and catering is done by a chef, engineering undertaken not as a side duty of one fisher but as three individual positions. But we also see a split between owners, managers and workers. It is hard to argue that these share a production unit on equal terms. We recognize the division between labor and owners as well as a manager role for the officers.

In the group of owners there is a passive investor, the skipper, the shore-based manager, and the upcoming skipper. There is also a group of producers consisting of a skipper and officers, engineers, fishers and a chef. Some of the owners are active on board while others are only engaged through their capital or land-based work. The same pattern was visible in the large company described in the example given in Chap. 5, and both companies used market-based fisheries management systems to expand and rationalize their production. This is a contemporary version of the capitalist mode of production in ocean-going fisheries. If we return to the opening question of why transferable fishing rights promote the large-scale capitalist fishery, part of the answer can be found in the simple commodity production and its adaptations and contractions when faced with the VQS market-based system; ultimately, it was unable to grow without increasing financial dependency. I have argued here and in Chapter 3 that large companies would gradually buy the majority of the VQS due to the financial aspect of the transfers. This promotes capitalist fisheries in the long run, but how does it change conditions in the short term?

The VQS and the Capitalist Mode of Production

Market-based fisheries management introduced a transferable commodity for the management of primary raw material, namely the fish resource, and it gave the rights holders the possibility to obtain loans using this new commodity as collateral. This enabled the producers to better plan their production and through financial means to expand this in order to operate at full scale. By monopolizing the resource (through individual ownership) it is protected against other potential users, who in turn are dependent on the owners for access. Better planning means that the risks for an unsuccessful fishing trip are lessened and that the costs of production can be reduced by targeting the fish in the right season. With the flexible leasing system, the risk of not catching the allocation can be counteracted. For a few fishing companies this has led to expansion, in which significant amounts of VQS were bought and accumulated from other operators. Historical rights from several vessels were transferred to fewer (most often larger) vessels. The conditions in the financial markets have an important role to play in this. When obtaining loans through banks the conditions are better for companies with investors, since extra security and capital input lead to a lower interest rate. Thus, capitalist companies can expand through advantages in the financial markets, even though they do not have better or more efficient fishing operations. In addition, the obtained VQS portions are not subject to local competition, since the resource is now the exclusive ownership of the company. There are, in other words, two effects of exclusive ownership: one related to production and the other to competition among operators.

As mentioned several times the VQS system introduces a new financial aspect to fishing activities. Two things are noteworthy in this respect. First of all, in many operations VQS has become the largest portion of the value. Second, due to its high value, the VQS alters the central aim of the fishing operation. That is, the operation is centered on the aim to harvest all of the allocated VQS—in other words, to pay off the capital. There has been a discrete displacement of the meaningful practice of fishing, from a wild-capture commodity production to a financial investment that has to be paid off. This was evident 1 year when the sand eel quota was reduced by 90 % and the operators complained about the loss of opportunity to pay off their investments in quota and vessels. They felt and expressed a right to a certain number of kilos for their production. This is a shift to what Marx would term M-C-M^I—from money to commodity to more money—where the meaning of production is to create more money than there was at the beginning. Why risk large amounts of money in the first place if it is not returned with a profit? This should be considered in contrast to the relation C-M-C, in which commodity producers produce (C) in order to obtain an income (M) to buy the other commodities (C) that are needed to make a living. The VQS itself is not enough for this change; but the financial aspect, capitalization of the resource and magnitude of the sums involved, both empower and force the owners to operate VQS as a commercial object in market-based fisheries management.

Crew and New Relations

We know that the share paid to the boat has been rising on the large and expanding vessels. From an individual point of view this has been made up by an increase in the annual turnover. Crew fishers on board receive a smaller share, but their income remains more or less constant because of the increase in production. At the individual level, this is acceptable; though it can become problematic on a societal scale or seen from the perspective of a labor union (Personal conversation, January 2012). The individual fisher receives a stable or even higher income, but there are fewer fishers in total. In the free and equal access system, vessel owners could not offer a radically different crew share than similar operations. But privatization changes this. No one could promise a better catch or end result unless the resource is already secured. This dynamic is changed by the introduction of market-based fisheries management. With ownership of quota shares the catch is defined and not subject to competition from other units potentially paying a better crew share. The bargaining power of the crew is weakened.

Seen from a job-creation perspective, less people are involved (sharing) and a larger share is returned to the capital. In this perspective people from numerous vessels are excluded and production organized in order to pay off capital, which is not only the investments in technology (vessel and gear) but now also in fishing rights. There are therefore a decreasing number of people employed in total; and with the resource under exclusive ownership, there are also fewer alternatives for the people inside the sector. In other words, decreasing boat shares are possible because unhappy "fish workers" have fewer other places to go, let alone the possibility to begin their own operation. This is a consequence of the new relationship between limited ownership of a resource and the agreement in the sector to use the share system as a pay structure. As we saw in the previous chapter, one operation under VQS had changed the pay structure to a combination of fixed wages and sharing, and this perhaps reflects these fundamental transformations.

Life-Modes of the Sea

The separation in ownership between VQS owners and non-owning workers also enables a landscape of multiple life-modes: investors, managers, and workers. With large shares of the resource as exclusive ownership there is need for managers to control and lead those who transform wild fish into landed fish. The allocated amount has to be harvested, and only by doing this efficiently can surplus value be produced and returned to the investors. Perhaps this is the meaning of the "efficient" fisher. A common interest from the crew perspective is to have the best conditions on board the vessel—to get something in return for the increased boat share. In the example above of Isafold, the new vessel has personal cabins with Internet and TV as well as an exercise room. A closer ethnographic study could examine how the good life as a wage laborer is lived partly on board the vessel and partly at home with family or friends—and how this group organizes their common interests. What are the workers' attitudes towards their work and "bosses"? How does the manager see his career on board the vessel? What is his or her relation to the (other) owners and the crew?

The introduction of market-based fisheries management has made it possible to organize fishing activities in a capitalist manner. In theory, the wage-worker, manager and investor all have different interests in the production unit. On one hand the capital investor might own the boat and company but know nothing of fishing; on the other hand the working fisher might have sophisticated knowledge about fish migration, for example, without profiting from a better catch. The boat and quota owner might also be considered a fisherman by the administration, although he rarely is on board.

Neither the statistical data nor economic theory provides insights into these complex relations; rather economic theory suggests they are only a result of the rational division of labor. However, these processes can be studied and described through qualitative fieldwork. In the new company described in Chap. 5, the crew was made up of workers from Poland. Migrating workers from Eastern Europe are a widespread phenomenon in Denmark, known as "east workers." Because of lower (but legal) salaries "east workers" substitute Danes in sectors with stiff competition and lots of manual work. In agriculture for example, it is estimated that 60 % of the workforce is foreign. In the following quote a skipper from a large company talks about the operations with "east workers" on board. Note the way he refers to the crew, including his own:

> I have the impression that those operations with east workers on board, they have to save money. They need the money for something else. I still make a large income despite having large wage expenses. For me it is related. There are some who have built boats and based their operation on the wage expense for only two men, and then they have hired Polish workers on the side to make ends meet. For me it does not make sense if you do not get anything for your fish. That is also what the auction leader says, those with east workers on board do not get much for their fish. They do not get the same price as those of us who put an effort in it. (Personal conversation, December 2011)

What is at stake is not shares but wage expenses. What is talked about is the way the production is organized in order to get a good price at the auction. The subjects in charge are the company owners and manager (the "We" and "I"), not a member of a shared organization. He continues:

> I have chosen from the beginning not to have east workers on board. We would rather go for quality, and they simply cannot deliver that. If we land with one of those operations we get much more. The Polish people do not care, they get 20,000 a month. If they put extra work into it, they still get 20,000 a month. If they piss or smoke and put the cigarette butts in the boxes they still get 20,000 a month. (Personal conversation, December 2011)

In other words the purpose of the pay structure in percentages is more about control and motivation of labor input than a shared production unit. It is an accord system, managed around Polish fish workers who have little interest in their product. Turning to the young, newly established skipper instead, we could ask why he does not join such an operation instead of putting so much effort into his own little operation. His answer is about independence:

> You run a routine in five years where you are in control, and do exactly what fits you, and then you work for another ship, where he says we will go and set the gillnets today, and then you say 'hell no', because that is not your plan. (Personal conversation, January 2012)

He is reluctant to work for others and not be in control himself. What he does, though, is occasionally to take a trip as a skipper on another vessel owned by a local:

> One of them has 230 days at sea a year, and sometimes he is tired of it [...] then he asks if I can take a trip with his ship. I have done it, it is not like that, but I also have my own business to run [...] then I get percentages, like the others, just some more of course. I got 14% and the others got 9–10%. I got 5% more because I have the responsibility. It might be you are just sitting in the wheelhouse, but you do more than the people on the deck. You have to set the nets and talk with the others, there is something all the time, write catch reports. (Personal conversation, January 2012)

Interestingly he talks about shares in his own operation and percentages when employed by someone else. The "career" work for others is just the exception, while his own business is his main objective. For the owner of the other vessel this arrangement only makes sense because he gets something in return. He is employing others to catch his allocation and pay off his investments. With the introduction of market-based fisheries management we are witnessing a much clearer division between the modes of production; with different types of work and providing different meanings to the diverging life-modes now engaged in the fisheries.

Conclusions

What I have shown in this chapter is that the introduction of market-based fisheries management, in several aspects, has changed the general conditions for the two modes of production. The introduction of private transferable fishing rights has transformed the commercial fisheries sector, to the benefit of fishing companies reorganizing and expanding their production. Through monopolization of the fish resource, large fishing companies can operate commercial enterprises in which a given quantity can be harvested and production planned in order to achieve full-scale operation and circulation of capital through the production unit. This is obtained through advantages in the financial system, and the new full-scale production level is the result of a new freedom for the captains of finance. However, it remains to be seen if full-scale production is not as much an illusion as it is a viable business strategy. The vessel can be expanded and other vessels employed, and thus expansion is only limited by the same thing that enabled it: the limited natural resource parceled out in quota shares. Can nature and technology deliver the large-scale advantages needed to uphold large-scale production? An equally important question is how investors and owners of large VQS shares will be able to exit the sector. If money capital and investments are to be returned as more money, who will be able to buy out the current large-scale owners? Will quota ownership be opened up for large corporate structures, for example fish processing companies? Will the already dubious legal border against foreign ownership of Danish marine resources be abandoned to allow international capital to facilitate the future distribution of fishing rights?

The private ownership of VQS has also led to changes in the most general conditions for share-organized fishers. The altered relationship between the vessel owner(s) and the remaining crew is a persistent challenge for this type of organization. For some this only becomes apparent when the operation is eventually sold and one fisher becomes jobless while the other is suddenly a millionaire. For others the boat share is changed, or leasing is a new cost paid by the crew. Only for the operators wholly without VQS is the share organization unaltered, as they are equal in regard to their lack of access. Besides the ownership structure, the daily and annual activities have to be planned. Here the owner suddenly has an amount of fish that must be caught, while the individual shares also allow for better planning. While the operation is still conceived of as a shared unit and operated as an equal partnership, the status between quota owners and non-quota owners has been changed.

Seen from the crew side, there is conflict over the best positions. Vessels with large amounts of VQS, with an effective, fair owner and reasonable boat shares are more attractive than others. Can we still argue that these fishing operations only enable self-employed fishers? This theoretical examination has highlighted how the new market-based fisheries management has made it possible to organize fishing activities in a capitalist manner. In large operations this was also present with a clear division between owners, managers and labor.

Seen in such a theoretical perspective, the fishing sector consists of a complex structure of life-modes and modes of production. These have diverse interests and objectives, as well as means and resources. If the policy framework is to provide the platform for *all* of these different life-modes, then at the political level we are faced with some urgent challenges. The distribution of fishing opportunities is marked by conditions in the financial sector, which promotes large companies. In the following chapter I sketch out a possible alternative to current market-based fisheries management. If there is political will to change this imbalance, the principles guiding the transfers will have to be changed in the near future. Instead, by constantly referring to the "fisher," managers and fisheries economists neglect and avoid the cultural and social aspect of the fishing practices. For the people involved, it matters they are wage-earners or if they are self-employed fishers.

References

Andersen, Raoul, and Cato Wadel. 1972. *North Atlantic fishermen: Anthropological essays on modern fishing*. Newfoundland social and economic papers, vol. no. 5. St. John's: Institute of Social and Economic Research, Memorial University of Newfoundland.

Andresen, Jesper, and Thomas Højrup. 2008. The tragedy of enclosure. The battle for maritime resources and life-modes in Europe. *Ethnologia Europaea* 38 (1): 29–41.

Frost, Hans, and Jørgen Løkkegaard. 2001. Individuelle omsættelige kvoter—Kort belysning af vigtige spørgsmål for danske fiskeri, ed. Fødevareøkonomisk Institut. Copenhagen: Statens Jordbrugs- og Fiskeriøkonomiske Institut.

Harvey, David. 2011. *The enigma of capital: And the crises of capitalism*. Pbk. Aufl. Oxford: Oxford University Press.

Hersoug, Bjørn. 2005. *Closing the commons: Norwegian fisheries from open access to private property*. Delft: Eburon.

Højrup, Thomas. 1983. *Det glemte folk: Livsformer og centraldirigering*. 3. opl. Aufl. Kbh.: Institut for Europæisk folkelivsforskning.

Højrup, Thomas. 2002. *Dannelsens Dialektik*. Copenhagen: Museum Tusculanums Forlag.

Højrup, Thomas. 2003. *State, culture and life-modes: The foundations of life-mode analysis*. Aldershot: Ashgate.

Löfgren, Orvar. 1977. *Fångstmän i industrisamhället: En halländsk kustbygds omvandling 1800–1970*. Lund: Liber Läromedel.

Monrad Hansen, Kirsten. 2012. European Fisheries at a Tipping Point. In *Estudios Europeos, No 1*, eds. Thomas Højrup, and Klaus Schriewer. Murcia: edit.um.

Resch, Robert Paul. 1992. Althusser and the renewal of Marxist social theory. Berkeley: University of California Press.

Tarbensen, Kenn. 2012. *For alle Danmarks fiskere: Danmarks Fiskeriforening 125 år, 1887–2012*. Fredericia: Danmarks Fiskeriforening.

Chapter 7
Postscript: Everyday Life and Mediated Fisheries

Abstract Two waypoints were identified at the beginning of this book. The first was a reflection on the different ways social sciences have conceptualized, criticized, and worked with market-based fisheries management. The second was a promise to show diversity and complexity in the social and cultural material. The two were related insofar as social diversity and cohesion are often emphasized by one branch of social science, particularly in the disciplines of ethnology and anthropology in response to more reductionist perspectives in economics and political science. In this book, I have argued that, in general, the two approaches had diverging views on market-based fisheries management, and I have suggested that these originate in the different research objects, instruments, and assumptions that underlie the social sciences. In this postscript, I reflect on the two waypoints, and I discuss the wider perspectives concerning the strong and international currents favoring market-based fisheries. In addition, I suggest *mediated fisheries* as a possible alternative management principle instead of distribution based purely on market mechanisms.

Keywords Fisheries economy · EU common fisheries policy · Developing countries · Privatization · Individual transferable quotas

The Economics of Ethnology

In my aim to describe the tensions between promoters and critics of market-based management, I have treated the disciplines of political economy and economics rather uniformly. Political economy and economy are fields with significant internal disciplinary differences, which of course prompt discussions of the basic assumptions and empirical consequences of these differences. I did however point at some fundamental assumptions that are characteristic of fisheries economics: namely the strong belief in individual rational choice and the focus on markets as providing a social good. This starting point yields a narrow definition of value as the rational product of exchanges between individuals, and consequently excludes other value systems. Fisheries economics have been dominating the management agenda in the last few decades and thus shaping the questions and solutions in their own ways. Perhaps it is time for a renewal of political economy as an influential discipline guiding a way through the complex ethics of managing limited resources. Central to this contribution is the argument that the social world of fisheries is diverse and made up of multiple ways of life that cannot be reduced to one basic human type or mode of operation.

A New Everyday

The introduction of market-based fisheries management is one of the most profound changes to the relationship between people and marine resources. By remaking this relationship, social relations and cultural meanings on land were also remade and changed. In Denmark, market-based fisheries management marked a substantial break with the principle of equal and free access to Danish marine resources, which were protected and regulated by the Danish state. The introduction of total allowable catches (TACs) and fishing quotas in the 1970s destabilized these principles. By privatizing these resources, the state gave away a common property it had been protecting for centuries against privatization and foreign fishing fleets. For some, the previous management system was an obstacle to realize a fully capitalist fishery, and a privileged group of vessel owners were empowered by the new system to engage in wide-scale expansion by monopolizing fish resources. This expansion was made possible by the use of the financial markets, with the new quota commodity as collateral. While it seems that some producers have managed to obtain full large-scale production, the economic performance of others indicates that in fishing the necessary conditions are not always present for large-scale production.

Expansionist companies run with rotating crews, absentee owners, and officers on board to manage a production that takes place year round in different catch areas. The social relationships between people on board these vessels resemble that of a factory. On the other hand, many share-organized fishers are seriously challenged by the introduction of private property rights, which alter the relationship between socially close colleagues. These fishers used to be the trademark of the Danish fisheries and formed highly competitive units driven by innovative captains of the industry. Historically, even with their relatively small units, fishers from north and west Jutland fished in large areas of the North Sea and outmatched other fleets in the area on quality and price of their catch. They relied on thrift, skills, and seasonal flexibility to adapt to the constantly changing conditions in the marine environment and on international markets. With the sudden change to market-based management this historical flexibility has been largely obstructed and altered, as different fish species are now privately-owned commodities. Today, these fishing units are increasingly dependent on leasing other people's fish, and as a whole this segment is in decline. Trawling companies are increasing their share of the activities while the numbers of gillnetters and Danish seiners are declining. Over time, this development includes an insight into the fundamental change that comes with market-based fisheries: the overall dynamic in this process is not related to who is the most efficient fisher but to who is willing to invest the most. The different conditions in the financial market are therefore decisive for future distribution, and in this system the self-employed fishers currently lose to the larger companies that promise investors a return on their capital investment. With the goal of being independent, many self-employed fishers hesitate and abstain from large investments and obligations through the financial market. On the other hand, the self-employed units have adapted. Some invested in quota shares (VQS) while others found niches in the leasing market that could provide a platform for a meaningful fishing operation (Fig. 7.1). Where the larger operators demand advantages of scale from a few species, there is room for the more flexible and labor-intensive units to focus on labor-intensive fisheries and some of the high-value species that are more difficult to catch.

But for all vessel owners, the capitalization of the resource has created new elements in their everyday practices. The markets for leasing and buying quota are constantly monitored and discussed, influencing both long and short-term strategies. With the quotas being many times more valuable than vessels and gear, difficult decisions influencing the crew, family, local community, and colleagues are dependent on fluctuations in a highly dynamic and unpredictable quota market. For those dependent on the leasing market, this has become a new factor like weather and fish migrations that from time to time provide a platform for a meaningful production. However, just behind leasing prices is a social relation marked by an uneven ownership of the resource. It is these everyday aspects of market-based fisheries management that I have shown and examined. I have inquired how the new commodity and market are conceptualized, integrated into existing social re-

Fig. 7.1 Both locals and tourists gather to see the fresh catch that the fisherman is bringing in. Will our generation be the last to witness this sight? (Photo: Jeppe Høst)

lations, and how the people involved have built their lives around the quota as a transferable commodity. Doing so, I have also tried to explain the new dynamics in a more theoretical manner and sought to identify general developments in the Danish context. It might be useful at this point to outline some of the wider perspectives in the research.

European Union and Beyond

Market-based fisheries management facilitates a power shift from communities and self-employed fishers to companies and the *captains of finance*, resulting in a loss of cultural diversity and opportunities for the inhabitants of local communities. To some degree it could be argued that it was necessary in a Danish context to restructure the fleet in relation to the limited available resource; and that market-based fisheries management was one of several ways to do this. As I described earlier, in relation to the reform of the European Union (EU) Common Fisheries Policy, the European Commission of Fisheries pushed for a model similar to the Danish one. The Commission's description of a well-balanced Danish model was—as I have shown—inaccurate. But even though at the time of writing Transferable Fishing Concessions (TFC) seem to be temporarily off the pan-European agenda as mandatory instruments, it is still up to each of the remaining countries in the EU to find a way to balance fleets and resources.

What this book has argued in both theory and practice is that market-based fisheries management should be seen as much more than a simple management tool. It is a profound change between people and a one-time solution with final and irreversible implications. Seeing market-based fisheries management as just a management instrument ignores and neglects the social, political, and economic context it is embedded in. On a global scale 200 million people depend in some part on small-scale fisheries for their livelihood (including some postharvest activities)[1]. Compared to the relatively well-organized Danes, how would communities and fishing sectors in developing countries react and transform under a market-based system? Little seems to indicate that market-based fisheries management would benefit and empower local communities and self-employed fishers in the developing world. Yet on the international scene there is a push for so-called clearly defined property rights in fisheries:

> The World Bank's PROFISH program recommends ending open access to fisheries and the single-minded competition for fish in favor of strengthening fishing rights for fishers. Well defined and secure fishing rights provide strong incentives to fishers, communities, and fishery associations to stop waste and overfishing, many believe.

---

[1] http://www.fao.org/fishery/fishcode-stf/activities/ssf/en (Accessed December 04, 2012)

Management systems that provide these kinds of rights, backed by force of law, are successfully used in Australia, Canada, Estonia, Greenland, Iceland, the Netherlands, New Zealand, and the United States. (Global Partnership for Oceans 2012: online document)[2]

Perhaps Denmark will soon be added to that list; though as I have shown in the case of the Danish VQS system, success is a subjective concept of an ideological nature, and a result of a fishing sector reshaped to fit a more capitalist empowered world. It is perhaps not surprising then to note the interest of the World Bank in the ideological and practical role of putting marine resources under capitalistic production and introducing market-based fisheries management in developing countries as well as in the rest of the world. Contrary to what powerful institutions such as the EU commission on fisheries and NGOs including the Environmental Defense Foundation claim, there is nothing in the Danish policy design to be exported or showcased for use in other countries. There is no intelligent social design; rather, naïve and insufficient management have resulted in an irreversible transformation of Danish fisheries.

Last Generation of Self-Employed Fishers

It has been a premise of this book that the social and political world consists of a complexity and diversity of social and cultural forms. Does the new market-based fisheries management cater to this complexity and allow multiple ways of life to exist in the fishing sector? It seems evident that larger fishing companies have gained power, and the numbers of self-employed fishers are declining. It would be tempting to propose that under the current policy we might be witnessing the last generation of self-employed fishers and that we are in the middle of a profound social and cultural transformation of the commercial fisheries. As I have shown, market-based fisheries management alters some of the most general conditions for the producers and favors large-scale capitalist mobile fisheries. With the resource as exclusive private property, it can be monopolized and operated for commercial purposes; a rent can be extracted and transformed into profit for a group of investors. At the same time, the permanent transfers of quotas seem to go to those with the best conditions on the financial market. As self-employed fishers seek to secure their independence, they are more reluctant to take on financial obligations. In this time of persistent financial and debt crises it seems contradictory that a policy design encourages economic risk-taking and punishes those with the most conservative attitude towards financial ventures. On the other hand, many self-employed fishers have found niches and methods to exist in the current system. Perhaps what we will see in the future is a constellation of large companies with substantial ownership of fishing rights existing alongside a diverse group of small holders using their flexibility of size while depending on the leasing market. If so, then the preceding chapters are an in-depth analysis of that transformation and therefore a description of how vessel owners and fishers, among others, have adapted to new politically-

---

[2] http://www.globalpartnershipforoceans.org/key-issues/giving-wild-fish-fighting-chance (Accessed December 04, 2012)

introduced conditions. But this research is also evidence that in the long run, financial conditions will redistribute the resource to the large companies with investor capital. As a final contribution, I will sketch out a simple idea of *mediated fisheries* that could overcome some of the challenges for young entrants and self-employed fishers in a market-based fisheries management system.

Mediated Fisheries

The idea of mediated fisheries is simple but relates to several of the findings presented in this book. In mediated fisheries, the transfers of fishing rights are directed by a producer community or simply the state. The basic idea is that every time a vessel owner wants to leave and sell his fishing rights, these can only be sold to the community or state (the mediator) at an annual fixed price. In turn the mediator will have to split the quota in equal shares and offer these to all of the remaining members. Three things follow from this idea. First of all, the price is fixed and therefore the chance to win as well as the risk of losing is greatly reduced. This will remove speculation and the casino character of the investments. Second, as the quota is redistributed in much smaller shares to the remaining operators, the financial obligations from investments are much more limited. Of course, operators should be able to abstain from buying what is offered; and what is left can be sold to those still interested or distributed in the next round. This might sound complicated, but it is close to the distributive principles in some of the quota pools, and the technology to overcome this on a large scale is already available. Nor does the complexity of a multispecies and multicatch area system challenge the basic principle—it only complicates it. Third, under this model the sector is financing generational handovers with no or little need of outside capital. This basic model could be further developed with designs directed at new entrants as well as crew fishers who could be allowed to buy shares.

This idea is of course based on similar principles to market-based fisheries management and does not articulate an alternative model for the initial distribution, as it emerged from an already existing system. However, the basic idea shows that principles of equality can be integrated into systems based on fixed allocations or boat shares. In that respect, the distributive model is embedded in either state organization or in a community of producers, more than solely on market mechanisms. The mediated fishery model draws on the findings in this study and is a contribution to the political and practical discussions on fisheries management. In other countries and regions, similar attempts to curb the market mechanisms are currently in place. In Norway, acquired fishing rights are in principle redistributed by the state after a period of 20 years (Hersoug 2005), and in Nova Scotia community quota groups were formed to manage the allocations (Sinclair et al. 1999). Challenges for the above sketch include how to define a community. In other words, who should be offered the VQS when an operator wants to leave, operators in a certain region or everyone? Could existing national and regional producer organizations be used as a framework? A similar predicament exists in regard to each species and its redistribution: should operators specialized in cod be offered plaice, and so on?

Private Fish and Captains of Finance

As the subtitle of this book indicates, the financial dimensions of running a fishing operation have gained radical importance. Not only are the investments in technology and vessel larger than ever before, but with the introduction of market-based fisheries management a whole new financial element has been introduced into fishing. I have been using the phrase captains of finance to emphasize how market-based fisheries management empowered a group of operators. There is however a double meaning to the phrase. By actively engaging in the financial expansion for production at full capacity, skippers and investors are not only doing this through finance, as captains in control of finance, but also increasingly as captains who are under the control of finance. Not only does the exchange-value become the decisive factor in planning the long-term strategy, the new status of the fish resource as capital also means that the value of fishing operations is subject to new uncontrollable factors and dynamics. In addition to weather, biology, and market prices of fish products, a new quota market means that the value of a fishing operation is taken out of the hands of the individual operators. The possibility for gain is accompanied by the risk of losing and being forced to run an operation at the mercy of a group of investors.

References

Hersoug, Bjørn. 2005. *Closing the commons: Norwegian fisheries from open access to private property*. Delft: Eburon.

Sinclair, M., R. O'Boyle, D. Burke, and F. Peacock. 1999. Groundfish management in transition within the Scotia-Fundy area of Canada. *Ices Journal of Marine Science* 56 (6).

MIX
Papier aus verantwortungsvollen Quellen
Paper from responsible sources
FSC® C105338

If you have any concerns about our products,
you can contact us on
ProductSafety@springernature.com

In case Publisher is established outside the EU,
the EU authorized representative is:
**Springer Nature Customer Service Center GmbH
Europaplatz 3, 69115 Heidelberg, Germany**

Printed by Libri Plureos GmbH
in Hamburg, Germany